THE
DRUNKEN
BOTANIST

ALSO BY AMY STEWART

From the Ground Up: The Story of a First Garden

The Earth Moved:
On the Remarkable Achievements of Earthworms

Flower Confidential:
The Good, the Bad, and the Beautiful

Wicked Plants:
The Weed that Killed Lincoln's Mother & Other Botanical Atrocities

Wicked Bugs:
The Louse that Conquered Napoleon's Army
& Other Diabolical Insects

The
DRUNKEN
BOTANIST

THE PLANTS THAT CREATE THE WORLD'S GREAT DRINKS

by
Amy Stewart

ALGONQUIN BOOKS OF CHAPEL HILL
2013

Published by
Algonquin Books of Chapel Hill
Post Office Box 2225
Chapel Hill, North Carolina 27515-2225

a division of
Workman Publishing
225 Varick Street
New York, New York 10014

Printed in China.
Published simultaneously in Canada by Thomas Allen & Son Limited.
Design by Tracy Sunrize Johnson.

Library of Congress Cataloging-in-Publication Data
   Stewart, Amy.
      The drunken botanist : the plants that create the world's great
   drinks / Amy Stewart.—1st ed.
         p. cm.
      Includes index.
      ISBN 978-1-61620-046-6
      1. Plants, Edible.   2. Plants, Useful.   3. Alcoholic beverages.
      4. Cocktails.   I. Title.
      QK98.5.A1S74 2013
      581.632—dc23   2012041725

20 19 18 17 16 15 14 13

TO PSB

contents

Aperitif x

About the Recipes xvi

PART I

WE EXPLORE THE TWIN ALCHEMICAL PROCESSES
OF FERMENTATION AND DISTILLATION,
FROM WHICH WINE, BEER, AND SPIRITS ISSUE FORTH

*Proceeding in an Orderly Fashion through the Alphabet:
The Classics, from Agave, 2, to Wheat, 107*

*Then Moving onto a Sampling of More Obscure Sources of Alcohol
from around the World: Strange Brews 111*

PART II

WE THEN SUFFUSE OUR CREATIONS
WITH A WONDROUS ASSORTMENT OF NATURE'S BOUNTY

*Herbs & Spices 135, Flowers 204, Trees 227,
Fruit 259, Nuts & Seeds 306*

PART III

AT LAST WE VENTURE INTO THE GARDEN,
WHERE WE ENCOUNTER A SEASONAL ARRAY OF
BOTANICAL MIXERS AND GARNISHES
TO BE INTRODUCED TO THE COCKTAIL IN ITS
FINAL STAGE OF PREPARATION

*Sorted in a Similar Fashion: Herbs 320, Flowers 327, Trees 332,
Berries & Vines 340, Fruits & Vegetables 345;
including Recipes and Sufficient Horticultural Instruction*

Digestif 355

*Some Final Business:
Recommended Reading 357, Acknowledgments 361, Index 363*

RECIPES

COCKTAILS

Classic Margarita (8)

The French Intervention (13)

Cider Cup (20)

The Vavilov Affair (25)

Rusty Nail (38)

Old-Fashioned (48)

Vermouth Cocktail (61)

Pisco Sour (67)

Black Gold (75)

No. 1 Sake Cocktail (80)

Manhattan (86)

Honey Drip (93)

Daiquiri (99)

Mojito y Mas (104)

Prickly Pear Sangria (128)

The Bay Rum (137)

Bison Grass Cocktail (147)

Pimm's Cup (161)

Dr. Struwe's Suze and Soda (164)

Moscow Mule (167)

The Classic Martini (173)

Dombey's Last Word (175)

The Perfect Pastis (181)

Sazerac (184)

Jerry Thomas' Regent's Punch (186)

Dancing with the Green Fairy (201)

The Aviation (226)

The Champagne Cocktail (229)

The Mamani Gin & Tonic (238)

The Douglas Expedition (244)

Royal Tannenbaum (251)

Caribou (258)

Valencia (261)

Kir (267)

The (Hybridized)
Brooklyn Cocktail (275)

Sloe Gin Fizz (283)

Red Lion Hybrid (289)

Negroni (292)

Ciao Bella (295)

The Frank Meyer Expedition (297)

Blood Orange Sidecar (301)

Ramos Gin Fizz (303)

Mai Tai (304)

Buena Vista's Irish Coffee (311)

Walker Percy's Mint Julep (326)

Lavender-Elderflower
Champagne Cocktail (330)

Lavender Martini (331)

Jack Rose (339)

The Frézier Affair (351)

Blushing Mary (354)

SYRUPS, INFUSIONS, AND GARNISHES

Prickly Pear Syrup (127)
Capillaire Syrup (185)
Elderflower Cordial (207)
Gomme Syrup (253)
Homemade Maraschino Cherries (273)
Homemade Nocino (316)
Garden-Infused Simple Syrup (324)
Brine Your Own Olives (336)
Homemade Grenadine (338)
Infused Vodkas (343)
Limoncello and Other Liqueurs (344)
Garden Cocktails: A Template for Experimentation (348)
Refrigerator Pickles (349)

THE INSPIRATION FOR THIS BOOK came from a chance encounter at a convention of garden writers in Portland, Oregon. I was sitting in a hotel lobby with Scott Calhoun, an agave and cactus expert from Tucson. Someone had just given him a bottle of Aviation, a fine locally made gin. "I'm not much of a gin drinker," he said. "I don't know what to do with this."

I knew what to do with it.

"I've got a drink recipe that will make you love gin," I said. He looked doubtful, but I continued. "We're going to need fresh jalapeños, some cilantro, a few cherry tomatoes . . ."

"Stop," he said. "That's enough. I'm in." No one from Tucson can resist a jalapeño-based cocktail.

We spent the afternoon running around Portland, gathering our ingredients. On the way, I subjected Scott to my rant on the many virtues of gin. "How can anyone with even a passing interest in botany not be fascinated by this stuff?" I said. "Look at the ingredients. Juniper! That's a conifer. Coriander, which is, of course, the fruit of a cilantro plant. All gins have citrus peel in them. This one has lavender buds, too. Gin is nothing but an alcohol extraction of all these crazy plants from around the world—tree bark and leaves and seeds and flowers and fruit." We had arrived at the liquor store by then, and I was gesturing wildly at the shelves around us. "This is horticulture! In all of these bottles!"

I hunted for the ingredient I needed—proper tonic water, made with actual cinchona bark and real *Saccharum officinarum*, not that artificial junk—while Scott browsed the selection of bottled *Agave tequilana*. He was in the habit of trekking into Mexico in search of rare agave and cactus, and he'd encountered many of his prized specimens coming out of the working end of a handmade Oaxacan still.

Before we left, we stood in the doorway for a minute and looked around us. There wasn't a bottle in the store that we couldn't assign a genus and species to. Bourbon? *Zea mays*, an overgrown grass. Absinthe? *Artemisia absinthium*, a much-misunderstood Mediterranean herb. Polish vodka? *Solanum tuberosum*—a nightshade, which is a weird family of plants if there ever was one. Beer? *Humulus lupulus*, a sticky climbing vine that happens to be a close cousin to cannabis. Suddenly we weren't in a liquor store anymore. We were in a fantastical greenhouse, the world's most exotic botanical garden, the sort of strange and overgrown conservatory we only encounter in our dreams.

The cocktail (Mamani Gin & Tonic, p. 238) was a hit with the garden writers. Scott and I signed copies of our books in our publisher's booth that night, and we took turns putting down our pens to slice peppers and muddle cilantro. The broad outlines of this book were conceived right then, over two or three of those decidedly botanical cocktails. I should dedicate this to the person who gave Scott that bottle of Aviation—if only either of us could remember who it was.

In the seventeenth century, British scientist Robert Boyle, one of the founders of modern chemistry, published his *Philosophical Works,* a three-volume treatise on physics, chemistry, medicine, and natural history. He understood perfectly the connection between booze and botany, which fascinates me as well. Here is an abridged version of his take on the subject:

> *The inhabitants of Carribbe islands supply us with remark-*
> *able instances hereof where the poisonous root Mandiboca*
> *is converted into both bread and drink: by being chew'd,*
> *and spit out into water, it soon purges itself of its noxious*
> *quality. They, having in some of our American plantations,*
> *found it very difficult to make good malt of maiz or Indian*
> *corn, they first reduce it to bread, and afterwards brew*
> *a very good drink from it. In China, they make their wine*
> *from barley; in the northern parts thereof, from rice and*
> *apples. In Japan, also they prepare a strong wine from rice.*
> *We in England, likewise, have great variety of wines from*
> *cherries, apples, pears, &c. little inferior to those of foreign*
> *growth. In Brazil, and elsewhere, they make strong wine*
> *of water and sugarcane: and in Barbadoes they have many*
> *liquors unknown to us. Among the Turks, where wine of the*
> *grape is forbid by their law, the Jews and Christians keep,*
> *in their taverns, a liquor made of fermented raisins. The*
> *Sura in the East-Indies is made of the juice that flows from*
> *the cocoa-tree; and sailors have often been inebriated, in*
> *that country, with the liquors made of the fermented juices*
> *obtain'd by the incision of vegetables.*

And so on. Around the world, it seems, there is not a tree or shrub or delicate wildflower that has not been harvested, brewed, and bottled. Every advance in botanical exploration or horticultural science brought with it a corresponding uptick in the quality of our spirituous liquors. Drunken botanists? Given the role they play in creating the world's great drinks, it's a wonder there are any sober botanists at all.

With this book, I hope to offer a plant's-eye perspective on booze and to supply a little history, a little horticulture, and even some agricultural advice for those of you who want to grow your own. I begin with the plants we actually turn into alcohol, such as grapes and apples, barley and rice, sugarcane and corn. Any of them can, with the help of yeast, be transformed into molecules of intoxicating ethyl alcohol. But that's only the beginning. A great gin or a fine French liqueur is flavored with innumerable herbs, seeds, and fruit, some of them added during distillation and some just before bottling. And once a bottle gets to the bar, a third round of plants are called into service: mixers like mint, lemon, and—if the party's at my house—fresh jalapeño. I structured the book around this journey from mash tub and still, to bottle, to glass. Within each section, the plants are arranged in alphabetical order by their common name.

It would be impossible to describe every plant that has ever flavored an alcoholic beverage. I am certain that at this very moment, a craft distiller in Brooklyn is plucking a weed from a crack in the sidewalk and wondering if it would make a good flavoring for a new line of bitters. Marc Wucher, an Alsatian eau-de-vie maker, once told a reporter, "We distill everything except our mothers-in-law," and if you've ever been to Alsace, you know he wasn't exaggerating.

So I was forced to pick and choose from the world's botanical bounty. Although I tried to cover some of the more obscure, exotic, and forgotten plants we imbibe, and to tell of some strange brews you'd have to travel the globe to sample, most of the plants you'll meet in this book will be familiar to American and European drinkers. I covered 160 in all and could have easily explored a few hundred more. Many of them have botanical, medicinal, and culinary histories so vast that a few pages can't do them justice—and in fact, some of them, such as quinine, sugarcane, apples, grapes, and corn have already received the book-length treatment they deserve. What I hope to do here is to give you just a taste of the dazzlingly rich, complex, and delicious lives of the plants that go into all those bottles behind the bar.

Before we proceed, a few disclaimers are in order. The history of drinking is riddled with legends, distortions, half-truths, and outright lies. I didn't think any field of study could be more prone to myths and misstatements than botany, but that was before I started researching cocktails. Facts tend to get bent out of shape over a round of drinks, and liquor companies aren't obligated to stick to the truth at all: their secret formulas can remain a secret, and the burlap bags of herbs placed about the distillery might be there only for ambience or even for misdirection. If I state plainly that a liqueur contains a particular herb, that's because the manufacturer or someone else with direct, firsthand knowledge of the process, said it did. Sometimes one can only guess at secret ingredients, so I've tried to make it clear when I'm guessing as well. And if the story of a beverage's origin seems dubious or cannot be verified from anything other than a single, yellowing newspaper clipping, I'll let you know that, too.

To those of you with more than a passing interest in distillation or mixology, I urge you to be wary of experimenting with unknown plants. As the author of a book on poisonous plants, I can tell you that dropping the wrong herb into a still or a bottle for the purpose of extracting its active ingredients might be your last act of creativity. I've included some warnings about deadly look-alikes and dangerous botanical relatives. Do remember that plants employ powerful chemicals as defenses against the very thing you want to do to them, which is to pluck them from the ground and devour them. Before you go foraging, get a reputable field guide and follow it closely.

It is also important to note that distillers can use sophisticated equipment to extract flavorings from a plant and leave the more harmful molecules behind, but an amateur soaking a handful of leaves in vodka has no such control. Some of the plants described in this book are poisonous, illegal, or tightly regulated. Just because a distiller can work with them safely doesn't mean you can, too. Some things are best left to the experts.

Finally, a word of caution about medicinal plants. The history of many of the herbs, spices, and fruits in this book is the very history of medicine. Many of them were traditionally used, and are still being used, to treat a range of ailments. I find that history fascinating and I've shared some of it here, but none of it is intended as medical advice. An Italian digestif can be surprisingly soothing to a troubled stomach or a troubled mind; beyond that, I'm unwilling to speculate.

Every great drink starts with a plant. If you're a gardener, I hope this book inspires a cocktail party. If you're a bartender, I hope you're persuaded to put up a greenhouse or at least plant a window box. I want everyone who walks through a botanical garden or hikes a mountain ridge to see not just greenery but the very elixir of life— the aqua vitae—that the plant world has given us. I've always found horticulture to be an agreeably intoxicating subject; I hope you will, too. *Cheers!*

ABOUT THE RECIPES

These are simple, classic recipes that best express the way a particular plant can be put to use in liquor. There are several original recipes, but even they are variations on the classics. If you're new to mixing drinks, here are a few hints.

SERVING SIZE: A cocktail is not supposed to be an enormous drink. The modern martini glass is a monstrosity; filled to the rim, it holds eight ounces of liquid. That's four to five drinks, more than anyone should choke down in a single sitting. (If nothing else, the liquor gets warm before you finish it.)

A serving of straight liquor is one and a half ounces, which is, conveniently, the larger end of a jigger. (The smaller side, called a pony, is three-quarters of an ounce.) Add liqueur or vermouth, and a not-too-excessive drink might contain the equivalent of two ounces of hard alcohol.

The recipes in this book conform to that standard. A nicely proportioned drink, sipped while it's still cold, is a lovely thing. Have a second one if you want, but do get in the habit of mixing one small, civilized drink at a time. To facilitate this, measure all your pours, and please get rid of your jumbo-sized cocktail glasses (or reserve them for drinks that are mostly fruit juice), and invest in a set of more modestly proportioned stemware. Oh, and speaking of glasses, for the recipes in this book, you can get by with Champagne flutes, wine glasses, and the following:

- *Old-Fashioned glasses*—Short, wide six- to eight-ounce tumblers.

- *Highball glasses*—Taller glasses that hold around twelve ounces. A standard sixteen-ounce drinking glass, or, for that matter, a Mason jar, will do.

- *Cocktail glasses*—Conical or bowl-shaped glasses with a stem; the basic martini glass.

ICE: Do not be timid about adding ice or a splash of water to a drink. It does not water down the drink; it improves it. Water actually loosens the hold that alcohol has on aromatic molecules, which heightens rather than dilutes the flavor.

MUDDLING: To muddle is to mash herbs or fruit in the bottom of a cocktail shaker, often with a blunt wooden implement called a muddler. If you don't have one of those, use a wooden spoon. Cocktails made with muddled ingredients are strained so that crushed plant matter doesn't end up in the glass.

SIMPLE SYRUP: Simple syrup is a mixture of equal parts water and sugar, heated to a boil to dissolve the sugar, then allowed to cool. Sugar water will attract bacteria, so don't bother mixing up a large batch—it won't keep long. Just mix a little when you need it. If time is short, a microwave and a freezer can speed up the boiling and subsequent cooling considerably.

STANDARD-ISSUE EGG WHITE WARNING: Some of the recipes call for raw egg whites. If you are concerned about the possible health consequences of consuming raw eggs, please skip those.

TONIC WATER: Don't ruin high-quality liquor with terrible tonic. Look for premium brands like Fever-Tree or Q Tonic, which are made with real ingredients, not artificial flavors and high-fructose corn syrup.

VISIT DRUNKENBOTANIST.COM
FOR MORE RECIPES AND TECHNIQUES.

PART I

*We Explore the Twin Alchemical Processes
of Fermentation and Distillation,
from which Wine, Beer, and Spirits Issue Forth*

The botanical world produces alcohol in abundance.
Or, to be more precise, plants make sugar, and when sugar
meets yeast, alcohol is born. Plants soak up carbon dioxide
and sunlight, convert it to sugar, and exhale oxygen.
It is not much of an exaggeration to claim that the very process
that gives us the raw ingredients for brandy and beer
is the same one that sustains life on the planet.

THE CLASSICS

WE BEGIN BY EXPLORING THE CLASSICS,
THE PLANTS MOST COMMONLY
TRANSFORMED INTO ALCOHOL,
PROCEEDING IN AN ORDERLY FASHION
THROUGH THE ALPHABET
from agave to wheat.

AGAVE

Agave tequilana
AGAVACEAE (AGAVE FAMILY)

The agave is better known for what it is not than for what it is. Some people think it is a kind of cactus; in fact, it is a member of the botanical order Asparagales, making it more similar to asparagus and a few other unlikely relatives: the shade-loving garden ornamental hosta, the blue hyacinth bulb, and the spiky desert yucca.

Another misconception arises when agaves are called century plants, suggesting that they bloom once in a hundred years. In fact, many bloom after eight to ten years but "decade plant" doesn't sound nearly as romantic. The much-anticipated bloom is vitally important, however: it yields the raw ingredients for tequila, mezcal, and dozens of other drinks distilled or fermented from this strange, heat-loving succulent.

pulque

The first drink to be made from agave was pulque, a mildly fermented beverage derived from the sap, or *aguamiel*. We know from remnants found at archeological digs that agave—called maguey in Mexico—was cultivated, roasted, and eaten eight thousand years ago; the sweet sap surely would have been drunk as well. Murals dating to 200 AD at the pyramid in Cholula, Mexico, depict people drinking pulque. The Aztec Codex Fejérváry-Mayer, one of the few pre-Columbian books not destroyed by the Spanish, portray Mayahuel, goddess of the agave, breast-feeding her drunken rabbit children, presumably offering them pulque instead of milk. She had four hundred children in all—the "Centzon Totochtin"—and they are known as the rabbit gods of pulque and intoxication.

The strangest bit of evidence for pulque's ancient origins comes from a botanist named Eric Callen who, in the 1950s, pioneered coprolite analysis, or the study of human feces found at archeological sites. He was ridiculed by his colleagues for his bizarre specialty, but he did make some astonishing finds concerning the diet of ancient people. He claimed that he could confirm the presence of "maguey beer" in two-thousand-year-old feces just from the odor of the rehydrated samples in his laboratory—which is either a testament to his sensitive nose or to the powerful bouquet of very old pulque.

To make pulque, the flowering stalk of the agave is cut just as it starts to form. The plant waits its entire life for this moment, stockpiling sugars for a decade or more in anticipation of the emergence of this single appendage. Cutting it forces the base to swell without growing taller; at that point, the wound is covered and allowed to rest for several months while the sap builds. Then it is punctured again, causing the heart to rot. This rotten interior is scooped out and the inside of the cavity is repeatedly scraped, which irritates the plant so much that sap begins to flow profusely. Once it begins flowing, the sap is extracted every day by means of a rubber tube or, in the old days, a pipette made from a gourd called *acocote.* (The *acocote,* in case you are inclined to grow your own, is often made from the long, skinny segment of *Lagenaria vulgaris,* a common bottle gourd also used to make bowls and musical instruments.)

A single agave can produce a gallon a day for months at a stretch, yielding over 250 gallons in all, far more than the plant would contain at any given time. Eventually the sap runs dry and the agave crumples and dies. (Agaves are monocarpic, meaning that they bloom only once and then expire, so this is not as much of a tragedy as it may seem.)

The sap needs less than a day to ferment—historically, this took place in wooden barrels, pigskins, or goatskins—and then it is ready to drink. A bit of the previous batch, the "mother," is usually added to start the process. It ferments quickly thanks in part to the naturally occurring bacteria *Zymomonas mobilis* that live on the agave and on other tropical plants that are made into alcohol, such as sugarcane, palms, and cacao. (These bacteria do such an efficient job of producing ethanol that they are used to make biofuels today.) However, this microbe is entirely unwelcome in other brewing processes. It is the cause of "cider sickness," a secondary fermentation that can ruin a batch of hard cider. It can spoil beer as well, releasing a nasty, sulfuric smell in a tainted batch. Still, it is the perfect catalyst for turning agave sap to pulque. *Saccharomyces cerevisiae,* the common brewing yeast, helps with fermentation, as does the bacterium *Leuconostoc mesenteroides,* which grows on vegetables and also ferments pickles and sauerkraut.

These and other microorganisms bring about a quick, frothy fermentation. Pulque is low in alcohol—only 4–6 percent alcohol by volume (ABV)—and has a slightly sour flavor, like pears or bananas past their prime. It is something of an acquired taste. Spanish historian Francisco López de Gómara, writing in the sixteenth century, said: "There are no dead dogs, nor a bomb, that can clear a path as well as the smell of [pulque]." Gómara might have preferred *pulque curado,* which is pulque flavored with coconut, strawberry, tamarind, pistachio, or other fruits.

Because no preservatives are added, pulque is always served fresh. The yeasts and bacteria remain active and the taste changes within a few days. Canned, pasteurized versions are available, but the microbes die off and the flavor suffers. It is, after all, the lively microbial mix that wins pulque comparisons to yogurt as well as beer. With its healthy dose of B vitamins, iron, and ascorbic acid, pulque is practically considered a health food. While beer has been the beverage of choice in Mexico for decades, pulque is making a comeback not only in Mexico but in border cities like San Diego as well.

mezcal and tequila

Any number of popular books on tequila and mezcal claim that when the Spanish arrived in Mexico, they needed a stronger drink to fortify themselves against the long and bloody struggle to come and introduced distillation as a way to turn pulque into a higher-proof spirit. In fact, tequila and mezcal are made from entirely different species of agave than pulque. The method for harvesting the plant and making the spirit is completely different, too.

It turns out to be very difficult to put pulque in a still and get strong liquor from it. The complex sugar molecules in agave nectar don't break down readily during fermentation, and heat from distillation causes unpleasant chemical reactions that create nasty flavors like sulfur and burning rubber. Extracting agave sugars for distillation requires a different technique—one that had already been perfected before the Spanish arrived.

Archeological evidence—including the aforementioned coprolite analysis carried out by Eric Callen and others—proves that people living in Mexico prior to the Spanish invasion enjoyed a long tradition of roasting the heart of the agave for food. Pottery fragments, early tools, paintings, and actual remnants of digested agave all confirm this beyond a doubt. Roasted agave is a gourmet experience; imagine a richer, meatier version of grilled artichoke hearts. It would have made a fine meal by itself.

But a high-proof spirit can also be made from the roasted hearts. The roasting process breaks down the sugars in a different way, yielding lovely caramelized flavors that make for a rich, smoky liquor. When the Spaniards arrived, they observed the locals tending to agave fields, monitoring the plants closely, and harvesting them at a precise point in their development, right before the bud emerged from the base to form a flowering stalk. Instead of scraping out the center to force the flow of sap, as was the practice for making pulque, the agave leaves were hacked away, revealing a dense mass called a *piña,* which resembled a pineapple or an artichoke heart. Those were harvested and roasted in brick or stone-lined ovens set in the ground, then covered so that they could smolder for several days.

MEZCAL *or* MESCAL?

While Americans and Europeans may prefer the spelling *mescal,* Mexicans always spell the name of their spirit *mezcal,* and that is its legal name under Mexican law.

THE LUMPERS, THE SPLITTERS, AND HOWARD SCOTT GENTRY

PERHAPS YOU'VE NEVER WANDERED THE MEXICAN DESERT WITH A FIELD GUIDE, attempting to identify the wild agaves growing there. It's not nearly as satisfying a pastime as, say, bird-watching: many agave species are nearly impossible to tell apart. Those that do look different might not in fact be so biologically distinct that they deserve to be classified as separate species—they might simply be different varieties. Think about tomatoes, for instance: a cherry and a beefsteak variety might not look or taste alike, but they are both members of the tomato species *Solanum lycopersicum*.

The same has happened with agaves. Howard Scott Gentry (1903-93) was the world's leading authority on agaves. As a plant explorer for the U.S. Department of Agriculture (USDA), he collected specimens in twenty-four countries. He believed that taxonomists (who are sometimes called lumpers or splitters for their tendency to either lump too many species together or separate too many varieties into distinct species) had done too much splitting in the case of agave. He argued that the difference between *A. tequilana* and other species was so insignificant that *A. tequilana* might not even be a separate species. He favored dividing the agaves by their floral characteristics, even though this would force botanists to wait as long as thirty years to see a specimen in bloom before they could properly identify it.

His colleagues Ana Valenzuela-Zapata and Gary Paul Nabhan, continuing his work after his death, have argued that a number of species, including *A. tequilana*, should, from a purely scientific perspective, be rolled into a broader species, *A. angustifolia*. But they acknowledge that history, culture, and the codification of *A. tequilana* in Mexican liquor laws make this difficult. Sometimes, tradition still trumps botany—especially in the Mexican desert.

Native people had clearly worked out a method for cultivating and roasting the agave. Pre-Columbian stone pits built for this purpose can still be found in Mexico and the southwestern United States. Now some archeologists point to remnants of crude stills to suggest that people might not have simply roasted the agave for food—they might have already been working on distillation methods prior to European contact.

This is a controversial idea hotly debated among academics. What we know for certain is that the Spaniards introduced new technology. Many of the earliest stills in Mexico are a derivation of the Filipino still, a wonderfully simple bit of equipment made entirely from local materials—mostly plants themselves. The reason the Spaniards get credit for this is that they are the ones who brought the Filipinos to Mexico, courtesy of the Manila-Acapulco galleons. These trading ships took advantage of favorable breezes that made it possible to journey directly from the Philippines to Acapulco in just four months' time. For 250 years, from 1565 to 1815, the ships brought spices, silk, and other luxuries from Asia to the New World, and they carried back Mexican silver for use as currency. The cross-pollination of cultures between Mexico and the Philippines survives even today, with the Filipino still being just one example of the connection between the two regions.

This simple still consisted of a hollowed-out tree trunk (often *Enterolobium cyclocarpum*, a tree in the pea family called guanacaste, or elephant ear) perched above an inground oven lined with bricks. The fermented mixture would be placed inside the tree trunk and brought to a boil. A shallow copper basin sat atop the tree trunk

WHAT ABOUT MESCALINE?

Mezcal is sometimes confused with mescaline, the psychoactive component of the peyote cactus *Lophophora williamsii*. In fact, the two are entirely unrelated, although peyote was sold in the nineteenth century as "muscale buttons," leading to a linguistic misunderstanding that persists today.

so that the liquid could boil and rise to the copper basin, much like steam collecting in the lid of a pot. This distilled liquid would then drip onto a wooden chute placed below the basin and run out of the still by way of a bamboo tube or a rolled agave leaf. More traditional copper Spanish stills, called Arabic stills, were also introduced early on.

Whenever distillation started in Latin America, the practice was well established by 1621, when a priest in Jalisco, Domingo Lázaro de Arregui, wrote that the roasted agave hearts yielded "a wine by distillation clearer than water and stronger than cane alcohol, and to their liking."

Over the last few centuries—and until the last decade or so—agave-based spirits were considered to be rough products that in no way compared to a good Scotch or Cognac. In 1897, a *Scientific American* reporter wrote that "mezcal is described as tasting like a mixture of gasoline, gin and electricity. Tequila is even worse, and is said to incite murder, riot and revolution."

While gin and electricity sound like excellent ingredients for a cocktail, this wasn't exactly a ringing endorsement. But today, artisanal distilleries in Jalisco and Oaxaca are making extraordinary smooth and fine spirits, using a mixture of ancient and modern technology.

CLASSIC MARGARITA

1½ ounce **tequila**
½ ounce **freshly squeezed lime juice**
½ ounce **Cointreau or another high-quality orange liqueur**
Dash of agave syrup or simple syrup
Slice of lime

Use good 100% agave tequila. A blanco would be the classic choice, but feel free to experiment with aged tequilas. Shake all the ingredients except the slice of lime over ice and serve straight up in a cocktail glass or over ice in an Old-Fashioned glass. Garnish with the slice of lime.

Mezcal at its best is a fine, handcrafted spirit, made in very small batches in Mexican villages using ancient techniques and a wide variety of wild agaves. The *piñas* are still chopped and roasted slowly in belowground ovens, where they are infused with the smoke from local oak, mesquite, or other wood for several days. They are then crushed by a stone wheel called a *tahona*. The wheel rolls around a circular pit, propelled in the old days by a donkey, although more sophisticated machinery is sometimes used today. (This wheel, by the way, is strikingly similar to apple-grinding stones once used to make cider in Europe. Whether the Spanish introduced the *tahona* to Mexico is a subject of hot debate among archeologists and historians.)

Once the roasted *piñas* are crushed, the juice can be siphoned off and fermented with water and wild yeast for a lighter-tasting mezcal, or the whole mash, including the crushed bits of agave, can be fermented, yielding a rich and smoky mezcal that would please any Scotch drinker. In some villages, the distillation takes place in a traditional clay and bamboo still. Other distillers use a slightly more modern copper pot still that is very similar to those used to make fine whiskies and brandies. Many mezcals are double- or triple-distilled to perfect the flavor.

Some distillers are so particular about their process that they won't let visitors near the still if they've used any perfumed soaps, fearing that even a few fragrance molecules will taint their product. The better mezcals are labeled by the species of agave and village, the way a good French wine would be. Today, according to Mexico's laws, a spirit carrying the name mezcal can only be made in Oaxaca and the adjacent state of Guerrero, and in three states to the north, Durango, San Luis Potosí, and Zacatecas.

There is one ingredient that can make mezcal different from whiskey or brandy: a dead chicken. *Pechuga* is a particularly rare and wonderful version of mezcal that includes wild local fruit added to the distillation for just a hint of sweetness, and a whole raw chicken breast, skinned and washed, hung in the still as the vapors pass over it. The chicken is supposed to balance the sweetness of the fruit. Whatever its purpose, it works: do not pass up an opportunity to taste *pechuga mezcal*.

What makes tequila different? For centuries, the term *mezcal* applied generally to all Mexican spirits made from the roasted heart of the agave. In the nineteenth century, *tequila* simply applied to mezcal made in or around the city of Tequila, in the state of Jalisco. It might have been made with a different species of agave, but the method was generally the same.

During the twentieth century, tequila settled into the drink it is today: a spirit made only in a designated area around Jalisco, from a cultivar of *Agave tequilana* called 'Weber Blue', often farmed in large fields rather than wild-harvested, and heated and steamed in an oven rather than slowly roasted in an underground pit. (Twenty-ton autoclaves are not an uncommon sight at tequila distilleries today.) Unfortunately, the definition of *tequila* also expanded to include *mixtos,* tequilas distilled from a mixture of agave and other sugars, with as much as 49 percent of the fermentation coming from non-agave sugar. Most tequilas Americans slurp down in the form of margaritas are *mixtos;* it still takes a little extra effort to order a 100% agave tequila. When you do, they are well worth sampling. Some are as sweet as an aged rum or as smoky and woodsy as a good whiskey, and some have unexpected floral notes, like a French liqueur. They are perfect on their own; there's no need to pollute a fine, handcrafted tequila with lime juice and salt.

Now that mezcal and tequila have their own appellation (called a DO, or Denominación de Origen in Mexico), other agave-based spirits are claiming their territory. *Raicilla* comes from the area around Puerto Vallarta, *bacanora* from Sonora, and *sotol*, made from the related desert spoon or sotol plant *Dasylirion wheeleri*, from Chihuahua.

protecting the plants

As these spirits become more popular, a new problem arises for Mexican distillers: protection of the plants and the land. Many of the non-tequila spirits are made from wild agaves. Some distillers of these spirits see the population of wild plants as being nearly unlimited and impossible to decimate; unfortunately, this is the same belief system that led to the destruction of the coast redwoods and other wild plant populations. Although some agaves reproduce vegetatively, producing "pups," offshoots that can regrow after

A FIELD GUIDE
TO TEQUILA AND MEZCAL

100% AGAVE: Must be made entirely from *A. tequilana* 'Weber Blue', in the DO, with no added sugars. Must be bottled by the producer in Mexico. May also be called *100% de agave, 100% puro de agave*, and so on. For mezcal, must be made from one of a number of approved agave species, in the DO, with no added sugars.

TEQUILA: A bottle simply labeled "tequila" is a *mixto*, meaning that it can be made with up to 49 percent non-agave sugars. It can be bottled outside the DO under certain conditions. Do yourself a favor and skip the *mixtos*.

SILVER (BLANCO OR PLATA): Unaged.

GOLD (JOVEN OR ORO): Unaged. For tequila, may be flavored and colored with caramel color, oak natural extract, glycerin, and/or sugar syrup.

AGED (REPOSADO): Aged in French oak or white oak barrels for at least two months.

EXTRA AGED (AÑEJO): Aged at least one year in six-hundred-liter or smaller French oak or white oak barrels.

ULTRA AGED (EXTRA AÑEJO): Aged at least three years in French oak or white oak casks of no more than 600 liters.

WHO PUT THE WEBER
IN "WEBER BLUE" AGAVE?

IF YOU READ ANY NUMBER OF POPULAR BOOKS ON TEQUILA
(or browse the boozier corners of the Internet), you may learn that *A. tequilana* was named by a German botanist called Franz Weber, who visited Mexico in the 1890s. However, botanical literature says otherwise. Botanists may disagree about where a plant should be placed on its family tree, or what it should be called, but they can usually agree on one thing: the person who first named and described a plant. The International Plant Names Index (IPNI) is a global collaborative effort among botanists to publish standard information about every named plant in the world. Each plant is listed by its scientific name, and in parentheses after that is the standard abbreviation for the botanist who described it.

Thanks to IPNI, we know that *A. tequilana* (F. A. C. Weber) was first described by Frédéric Albert Constantin Weber in an article published in a Parisian natural history journal in 1902. From his obituary, published when he died in 1903, we know that he was born in Alsace, completed his training as a doctor of medicine in 1852, published his thesis on the subject of cerebral hemorrhage, and promptly joined the French military, where his skills certainly would have been put to use. He was sent to Mexico just as the French, under the command of Napoleon III, joined Britain and Spain in invading Mexico to collect on unpaid debts. The short-lived imposition of the Austrian emperor Maximilian I, and his subsequent execution by firing squad, would not have left Dr. Weber with much time with which to indulge his hobby of plant collecting. Still, he managed to acquire and describe a number of cacti and agaves, which he cataloged in botanical journals after his return to Paris. Late in life, he served as the president of the Société Nationale d'Acclimatation de France, a nature conservation society. When his colleagues, writing in 1900, named *A. weberi* after him, they described his time in Mexico in even more detail, confirming that he was there in his official capacity—and collecting plants in his spare time—in 1866 and 1867.

So what about Franz Weber? If there was a German botanist by that name working in Mexico in the 1890s, his name has not been attached to a single plant in the scientific literature— and he certainly can't claim credit for naming *A. tequilana*.

THE FRENCH INTERVENTION

While most mezcal distillers are puzzled by the idea of mixing their spirit into a cocktail, American bartenders can't resist experimentation. In fact, tequila and mezcal both work beautifully in any cocktail that calls for whiskey, rye, or bourbon. This blend of French and Mexican ingredients is named after the 1862 French invasion of Mexico that brought Dr. Weber, who named *A. tequilana*, to the country.

1½ ounces *reposado* tequila or mezcal
¾ ounce **Lillet blanc**
Dash of green Chartreuse
Grapefruit peel

Shake all the ingredients except the grapefruit peel over ice and strain into a cocktail glass. Garnish with the grapefruit peel.

harvest, the harvest process prevents them from blooming. By not allowing the plants to flower, reproduce, and set seed, the genetic diversity is seriously impacted. Even the population of wild bats that pollinate agaves are diminished because the agaves are not allowed to bloom naturally.

The situation is worse for tequila, which generally comes from plants that have been farmed rather than harvested in the wild. Since only one species, *A. tequilana*, can be used to make the spirit, it has become a monoculture just as grapes have in northern California. David Suro-Piñera, owner of Siembra Azul tequila and an advocate for the preservation of tequila's history and the sustainability of the industry, said, "We've been abusing the species. We have not allowed the plant to reproduce in the wild. Genetically, it is exhausted and very vulnerable to disease. I'm very concerned." He attributes an increased use of pesticides, fungicides, and herbicides to the weakness of the plants themselves. Also, water is an important ingredient in tequila and other spirits; increased chemical use and degradation of the soil can pollute water supplies as well.

Already plagues of disease have devastated the domesticated agave crop, not unlike the catastrophic Irish potato famine or the wave of phylloxera that destroyed European vineyards. In the case of the agave, the agave snout weevil (*Scyphophorus acupunctatus*) introduces bacteria and deposits eggs that hatch into tiny larvae that eat the plant, rotting it from the inside out. Because the weevil bores inside, insecticides are largely ineffective.

Strengthening the crops and preserving wild agaves will require a combination of intercropping—the practice of interspersing agaves with other plants—protecting wild areas to increase genetic diversity, reducing chemical use, and taking steps to restore the health of the soil.

HOW TO TASTE

A fine tequila or mezcal should be savored on its own, in an Old-Fashioned glass, perhaps with a splash of water or a chunk of ice, just as you might drink a good whiskey. Lime and salt are unnecessary; their only purpose is to cover the taste of poor-quality spirits.

A SELECTED LIST OF AGAVES
and AGAVE-BASED SPIRITS

NOT ALL AGAVES ARE CREATED EQUAL. Some yield more sap and are better suited to the production of pulque, whereas others produce the kind of rich, fibrous heart that is perfect for roasting and distilling. Many species of agave are not used at all because they contain toxins and saponins, which are foamy, soaplike compounds that have steroidal and hormonal properties that make them unsafe to consume. Here are just a few that have been used, some for thousands of years:

AGAVA	*A. tequilana* (made in South Africa)
BACANORA	*A. angustifolia*
100% BLUE AGAVE SPIRITS	*A. tequilana* (made in the United States)
LICOR DE COCUY	*A. cocui* (made in Venezuela)
MEZCAL	By law, only the following can be used: *A. angustifolia* (*maguey espadín*), *A. asperrima* (*maguey de cerro, bruto o cenizo*), *A. weberi* (*maguey de mezcal*), *A. potatorum* (Tobalá), *A. salmiana* (*maguey verde o mezcalero*). Other agaves not already designated for use in another beverage under another DO in the same state may also be used.
PULQUE	*A. salmiana* (syn. *A. quiotifera*), *A. americana*, *A. weberi*, *A. complicate*, *A. gracilipes*, *A. melliflua*, *A. crassispina*, *A. atrovirens*, *A. ferox*, *A. mapisaga*, *A. hookeri*
RAICILLA	*A. lechuguilla*, *A. inaequidens*, *A. angustifolia*
SOTOL	*D. wheeleri* (an agave relative called the desert spoon)
TEQUILA	By law, only *A. tequilana* 'Weber Blue' can be used.

BUGS *in* BOOZE: WHAT ABOUT THE WORM?

The worm, or *gusano,* sometimes found at the bottom of a bottle of mezcal is the larva of the agave snout weevil (*S. acupunctatus*) or the agave moth (*Comadia redtenbacheri*)—and typically not, as is widely reported in the booze literature, *Hypopta agavis,* a moth that does feed on agave but causes less harm.

These grubs are added only as a marketing gimmick and are not a traditional part of the recipe. They are usually a sign of a cheap mezcal aimed at drinkers who don't know better. Makers of fine mezcal have lobbied, unsuccessfully, to have the worm banned entirely because they feel it denigrates the entire category. While the worm may not have an obvious influence over the flavor of mezcal, a 2010 study showed that the DNA from the larva was present in the mezcal it was bottled with, proving that *mezcal con gusano* does deliver a little bit of worm with every sip.

Another unfortunate marketing ploy is the addition of a scorpion, with its stinger removed, into a bottle of mezcal. Fortunately, the regulatory council governing tequila does not allow such nonsense in its bottles.

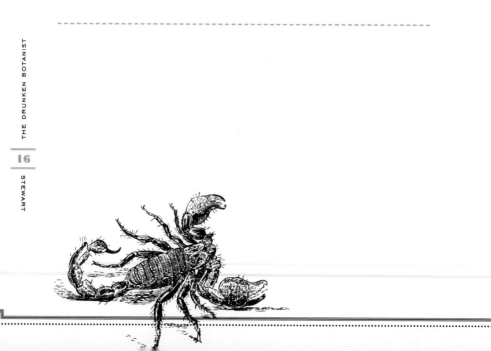

APPLE

Malus domestica
ROSACEAE (ROSE FAMILY)

The apple best suited for cider and brandy is what we would call a spitter: a fruit so bitter and tannic that one's first instinct is to spit it out and look around for something sweet to coat the tongue—a root beer, a cupcake, anything. Imagine biting into a soft green walnut, an unripe persimmon, or a handful of pencil shavings. That's a spitter at its worst. How, then, did anyone discover that something as crisp and bright as cider, or as warm and smooth as Calvados, could be coaxed from it?

The answer lies in the strange genetics of the apple tree. The DNA of apples is more complex than ours; a recent sequencing of the Golden Delicious genome uncovered fifty-seven thousand genes, more than twice as many as the twenty thousand to twenty-five thousand that humans possess. Our own genetic diversity ensures that our children will all be somewhat unique—never an exact copy of their parents but bearing some resemblance to the rest of the family. Apples display "extreme heterozygosity," meaning that they produce offspring that look nothing like their parents. Plant an apple seed, wait a few decades, and you'll get a tree bearing fruit that looks and tastes entirely different from its parent. In fact, the fruit from one seedling will be, genetically speaking, unlike any other apple ever grown, at any time, anywhere in the world.

Now consider the fact that apples have been around for fifty million to sixty-five million years, emerging right around the time dinosaurs went extinct and primates made their first appearance. For millions of years, the trees reproduced without any human interference, combining and recombining those intricately complex genes the way a gambler rolls the dice. When primates—and later, early humans—encountered a new apple tree and bit into its fruit, they never knew what they were going to get. Fortunately, our ancestors figured out that even bad apples make great liquor.

GROW your OWN

APPLES

SELECTING A TREE: A good fruit tree nursery will carry a selection of "cider apples" and will offer advice on choosing the right apple for any climate. Different apple cultivars require a different number of "chill hours"—the number of hours between November and February below 45 degrees—to break dormancy, so matching the tree to local winter weather conditions is important. Nurseries will also know whether a tree requires another nearby tree for cross-pollination; not all cultivars do.

ROOTSTOCK: Apples trees are grafted to rootstock that will control the growth of the tree, regulate production, and resist disease. M9 is a popular dwarf rootstock, allowing trees to reach only about ten feet in height. EMLA 7 reaches fifteen feet.

THINNING AND PRUNING: Cider apples tend to go biennial (meaning that they bear fruit every other year) if they are not thinned. Large orchards spray chemicals on apple blossoms after most of the flowers have opened, which will kill the open blossoms and significantly reduce the number that set fruit. Home gardeners simply pick a few apples from each cluster when the fruit is about the size of a grape. Ask a nursery or county extension office for pruning and thinning advice; they may offer workshops as well.

PESTICIDES: One of the great advantages of cider apples is that the trees naturally resist pests, and if they do experience a little damage from bugs, it matters less because the fruit is just going to be crushed anyway.

FULL SUN

DEEP INFREQUENT WATER

HARDY TO -25F/-32C

cider

The first boozy concoction to come from apples was cider. Americans refer to unfiltered apple juice as apple cider and usually drink it hot with a cinnamon stick. But ask for cider in other parts of the world and you'll get something far better: a drink as dry and bubbly as Champagne and as cold and refreshing as beer. When we drink it at all in North America, we call it hard cider to distinguish it from the nonalcoholic version, but such a distinction isn't necessary elsewhere.

The Greeks and Romans mastered the art of cider making. When Romans invaded England around 55 BC, they found that cider was already being enjoyed by the locals there. By that time, apple trees had long ago migrated from forests around Kazakhstan and were well established across Europe and Asia. It was in southern England, France, and Spain that the technique of fermenting—and later distilling—the fruit was perfected. Evidence of this ancient art can be found in the European countryside today, where large circular apple grinding stones used to crush the fruit are still half buried in the fields.

Because the oldest orchards were seedling orchards—meaning that every tree was started from seed, resulting in a mishmash of novel and never-before-seen apples—early cider would have been made from a blend of all the fruit in the orchard not sweet enough

PRESERVING HERITAGE CIDER APPLES

Keeping the world's great cider varieties alive is no simple matter. During World War I, the front line in the battle between the German and Allied forces happened to run right through Simon Louis-Frères' famous apple nursery near Metz, France. The 1943 Battle of Kursk devastated a thriving nursery and orchard trade south of Moscow. Today pomologists at Cornell University preserve strains in orchards in upstate New York as part of a global movement to catalog and save old apple varieties.

CIDER CUP

In the Middle Ages, people made a crudely fermented drink called *dépense* by steeping apples and other fruit in water and letting the juice ferment naturally. This is a much more refined version that is light enough to drink all afternoon in the summer.

2 parts hard cider
Sliced apples, oranges, melons, or other seasonal fruit
Frozen raspberries, strawberries, or grapes
1 part ginger beer or ginger ale (nonalcoholic)

In a large pitcher, combine the cider and sliced fruit; allow to soak for 3 to 6 hours. Strain to remove the sliced fruit. Fill highball glasses with ice and frozen berries, fill the glass three-quarters full with cider, and top with ginger beer to taste.

to eat. The only way to reproduce a popular apple cultivar was to graft it onto the rootstock of another tree, a technique that had been used on and off since 50 BC. Apple farmers started making clones through grafting, and those popular varieties eventually acquired names. In the late 1500s, there were at least sixty-five named apples in Normandy. For centuries, many of the best apples for cider-making have come from this region, all chosen for their productivity as well as their balance of acidity, tannin, aromatics, and sweetness.

In America, the toss of the genetic dice continued, with John Chapman, a man we know as Johnny Appleseed, establishing apple nurseries at the edge of the frontier in the early nineteenth century. He considered it wicked to start a tree by grafting, so his always grew from seed, the way nature intended. That means that early settlers grew—and made cider from—uniquely American apples, not the well-established English and French cultivars being grown across the Atlantic.

Historians love to trot out statistics on cider consumption before the twentieth century to demonstrate what lushes our ancestors were. In apple-growing regions, people drank a pint or more per day—but they had few alternatives. Water was not to be trusted as a beverage: it carried cholera, typhoid fever, dysentery, *E. coli*, and a host of other nasty parasites and diseases, many of them not well understood at the time but clearly originating in water. A mildly alcoholic drink like cider was inhospitable to bacteria, could be stored for short periods, and was safe and pleasant to drink, even at breakfast. Everyone drank it, including children.

Cider has always been low in alcohol because the apples themselves are low in sugar. Even the sweetest apples contain much less sugar than grapes, for instance. In a vat of cider, the yeast eat what sugar there is, turning it into alcohol and carbon dioxide, but once the sugar is gone, the yeast die off for lack of food, leaving behind a fermented cider that contains only about 4 to 6 percent alcohol.

Today, some cider makers bottle their product and then add another round of sugar and yeast, allowing the carbon dioxide to build up inside the bottle and create bubbles, Champagne-style. On the other

CLASSIFICATION OF APPLES
for CIDER MAKING

BITTERSHARP: Higher tannin, higher acidity (Kingston Black, Stoke Red, Foxwhelp)

SHARP: Low tannin, higher acidity (Granny Smith, Brown's, Golden Harvey)

SWEET: Low tannin, low acidity (Golden Delicious, Binet Rouge, Wickson)

BITTERSWEET: Higher tannin, lower acidity (Royal Jersey, Dabinett, Muscadet de Dieppe)

end of the spectrum, so-called industrial ciders made by large scale commercial distilleries may also contain non-fermenting sweeteners like saccharine or aspartame to give cider the sweetness that the mass market demands.

calvados and applejack

But there's more to apples than cider. In 1555, a Frenchman named Gilles de Gouberville wrote in his diary that a visitor had suggested a way to make a clear, highly alcoholic spirit from cider. Once fermented, he explained, cider could be heated, so that the alcohol would rise with the steam and collect in a copper pot, where it could be extracted and bottled. A little time in an oak barrel made it even better. The term for this spirit might have originally been *eau-de-vie*

APPLE SPIRITS

APPLE BRANDY: A generic term for a spirit distilled from fermented apple juice or mashed apples, bottled at a minimum of 40 percent ABV, usually aged in oak.

APPLEJACK: In the United States, another term for apple brandy. "Blended applejack" contains at least 20 percent applejack; the rest is neutral spirits.

APPLE LIQUEUR: A sweeter, lower-alcohol apéritif (often about 20 percent ABV) can be made from apples in a number of ways. One method would be to add apple brandy to fermenting cider before the yeast have consumed all the sugar. The higher alcohol content kills the yeast, stopping fermentation and resulting in a sweet drink almost like a dessert wine with fresh apple flavor. Apple liqueurs may be aged in oak before bottling.

APPLE WINE: While *apple wine* is a very old term for cider, today it refers to a type of cider to which additional sugars and yeasts have been added to push the alcohol content higher, usually to at least 7 percent ABV. Apple wines are typically not carbonated.

CALVADOS: Apple brandy made in a specific region of northern France, using apples from designated orchards, containing at least 20 percent local varieties, at least 70 percent bitter or bittersweet varieties, and no more than 15 percent sharp varieties. The spirit is bottled at a minimum of 40 percent ABV.

CALVADOS DOMFRONTAIS: Follows the other rules for Calvados, but this apple brandy must contain at least 30 percent pears. It is single distilled in a column still and aged in oak for at least three years.

CALVADOS PAYS D'AUGE: This is specific to the Pays d'Auge region; it follows all other rules for Calvados and must be double-distilled in a traditional copper still and aged in oak for at least two years.

EAU-DE-VIE: A clear spirit made from fermented fruits that is not aged in oak and is bottled at 40 percent ABV or higher. It is the fruit equivalent of "white whiskey."

POMMEAU: A delightful French blend of unfermented cider and apple brandy bottled at about 16 to 18 percent ABV.

de cidre—eau-de-vie being the early term for any kind of distilled spirit—but it soon earned the name Calvados, after the region in Normandy where it was made.

Americans wasted no time making their own version of Calvados. The Laird & Company Distillery in New Jersey holds bragging rights to License No. 1, the first distillery license issued in the United States, in 1780. According to the family's records, Alexander Laird arrived from Scotland in 1698 and began growing apples and making "cyder spirits," or applejack, for his friends and neighbors. When Robert Laird went to fight under George Washington's command, the family sent a gift of applejack for the troops. The family claims that Washington liked it enough to request the recipe and begin producing it on his own farm, but there is no record of applejack distillation at Mount Vernon. Cider, however, was regularly made for the Washington family, staff, and slaves.

Colonists who lacked the technical skills to build a copper still found another way to do it—they'd leave a barrel of cider outside in winter, let the water content freeze, and siphon off the unfrozen alcohol. The "freeze distillation" method was dangerous: with no way to extract the concentrated toxic compounds that can usually be removed during distillation, the alcohol contained enough poison to contaminate the liver or cause blindness. That might have given applejack an undeservedly bad reputation, but fortunately, better distillation methods prevailed.

Apples also make a fine eau-de-vie. Rather than running fermented apple juice through a still, eau-de-vie is typically made by crushing whole apples into a mash, fermenting it, and distilling a high-proof, clear alcohol. According to Cornell pomologist Ian Merwin, using whole crushed apples yields much higher levels of the aromatics that give apple spirits their flavor. "A good eau-de-vie made with mash fermentation tastes much more like an apple than Calvados does," he said. It also helps that it is usually distilled in a more sophisticated fractional column still, which allows for more precise retention of aromatics. Calvados, by French law, must be distilled in an older-style alembic pot still, which is a more traditional but less exacting method of distillation.

Eaux-de-vie are not finished in barrels, which means that the flavor comes entirely from the fruit and not from the oak. "With Calvados," Merwin said, "you're really just taking apple-based ethanol, which is a solvent, and putting it into oak to extract the oak flavors from it—which are admittedly nice in their own right. But there's not as much apple flavor left when it comes out of the barrel."

Don't tell that to a Calvados enthusiast. A nicely aged Calvados possesses a certain golden, sunlit quality that can only come from apples. It is best enjoyed neat, before or after dinner, or even in the middle of a meal: in Normandy the phrase trou normand, or "Norman hole," refers to the glass of Calvados served between courses to create a hole in the appetite and make room for the rest of the meal.

THE VAVILOV AFFAIR

Russian botanist Nikolai Vavilov risked everything to preserve the wild ancestors of the apple tree. In the early twentieth century, he traveled the world to identify the geographic origins of such important crops as apples, wheat, corn, and other grains, collecting seed from hundreds of thousands of plants to establish a seed bank and advance the science of genetics. His goal was to improve crop yields for Russian farmers, but Joseph Stalin considered him an enemy of the state. Stalin had some funny ideas about science: he believed that a person's behavior could change their genetic makeup, so that habits learned in one lifetime could be passed on through their DNA. Scientists who disagreed went to jail for it.

Vavilov was arrested for his beliefs in 1940. He spent his last days delivering lectures on genetics to the other prisoners, many of whom surely wished Stalin would have arrested some locksmiths or dynamite experts instead of botanists.

This version of an Old-Fashioned is mixed with equal parts applejack and bourbon, combining apples, corn, and grains in Vavilov's honor.

1 sugar cube
2 dashes Angostura bitters
¾ ounce applejack
¾ ounce bourbon
2 slices sharp apple, such as Granny Smith or Fuji

Place the sugar cube in the bottom of an Old-Fashioned glass. Splash the bitters and a few drops of water on the cube, and muddle. Add ice, the applejack, and the bourbon and stir well. Use a citrus squeezer to squeeze the juice of 1 apple slice on top. Add the second slice to the glass as garnish.

⊰ YEAST ⊱

- - - **(A LOVE STORY)** - - -

Saccharomycetales spp.

THE OLDEST DOMESTICATED LIVING ORGANISM IS NOT A HORSE OR A CHICKEN, nor is it corn or wheat. It is a wild single-celled, asexual creature capable of preserving food, making bread rise, and fermenting drinks. It is yeast.

Yeast is everywhere. It floats through the air, it lives on and inside of us, and it coats the skin of fruit in hopes of extracting some sugar from it. There's no need to go hunting for wild yeast—leave a bowl of flour and water on the kitchen counter, and yeast will find it. But a few particular species of yeast—especially those belonging to the *Saccharomycetales* genus—are so effective at fermentation that people learned to keep them alive, grow them in large quantity, and, eventually, to sell them to brewers and distillers. There are laboratories all over the world carefully tending their strains of yeast. Wineries, breweries, and distilleries are often reluctant to remodel, move, or replace equipment for fear of destroying the native yeast that have taken up residence and added their unique characteristics to their products. Tests on identical batches of apple cider show that the particular strain of yeast can radically influence the flavor, introducing unique fruit and floral notes to the finished brew.

The science of fermentation is wonderfully simple. Yeast eat sugar. They leave behind two waste products, ethyl alcohol and carbon dioxide. If we were being honest, we would admit that what a liquor store sells is, chemically speaking, little more than the litter boxes of millions of domesticated yeast organisms, wrapped up in pretty bottles with fancy price tags.

But as waste products go, those of yeast are endlessly useful. We'll dispense with the carbon dioxide first. If the fermentation is taking place in a vat, the carbon dioxide simply escapes. Beer makers allow some to remain so that the beer will be foamy. They might add a little back in during the bottling phase as well. In the case of sparkling wines, another bit of yeast goes into the bottle for a secondary fermentation that creates bubbles and builds up pressure behind the cork. (Bakers have much in common with brewers: carbon dioxide is what forces bread dough to rise.)

But what about the other waste product, ethyl alcohol? That is what we call pure alcohol, or ethanol. After some tinkering, it makes for a great drink—but not for the yeast. As they excrete this alcohol, yeast make their own grave. They can't survive in high concentrations of their own waste product, so as the alcohol content rises above about 15 percent, the yeast die off. That explains why, until distillation was invented, no human had ever enjoyed a stronger drink than beer or wine.

So that's how it ends for yeast. Either they run out of sugar and die of starvation, or they eat so much sugar that the alcohol they produce kills them. Either way, they die doing what they do best: making drinks for us.

- -

If ethanol were the only product excreted by yeast in a vat of fermenting sugars, the world's brandy makers and vodka distillers would have an incredibly easy time of it. They would simply dilute, flavor, and bottle the ethanol. But yeast, being living organisms, are imperfect, and the crushed grapes or mashed apples they live in are themselves brilliantly imperfect and complicated. There is more than sugar in a vat of grapes: tannins, aromatic compounds, acids, and forms of sugar that yeast cannot digest (also called non-fermenting sugars) are bouncing around as well. With so much going on in a fermentation tank, mistakes are going to happen.

Many of those "mistakes" take place as the enzymes inside yeast cells try to do their job, which is to regulate chemical reactions. Think of an enzyme as a lock in search of a key. As molecules jump around in the fermentation tank, they may try to "lock" with an enzyme but not quite fit. The result of these imperfect couplings are imperfect compounds—and these make fermented drinks complex, intricate, and sometimes dangerous.

These accidental by-products are called congeners—like the word *congenital*, meaning that the compounds were present from the birth of the fermented drink. Some of them are quite toxic and have to be carefully removed during distillation.

If those poisons are made during fermentation, why isn't beer or wine a deadly drink? First, brewers can control the fermentation process through their choice of equipment, the particular strains of yeast they use, and the temperature at which they allow fermentation to take place. Storing the fermented drink, or aging it in oak casks as winemakers do, causes further chemical reactions that can break down some compounds.

Some congeners inevitably remain, but they are present in such relatively small quantities that our livers are usually able to keep up. Anyone who has had too much wine has experienced a hangover caused, in part, by a buildup of these toxins that the body simply can't eliminate fast enough.

The challenge of distillation, then, is to extract ethyl alcohol from a fermented mash similar to beer or wine, resulting in a higher-alcohol spirit that does not also deliver a concentrated dose of congeners. Fortunately, each of these compounds has a different boiling point, so the secret is to heat the mixture and separate out the unwanted molecules as they boil away.

Light a fire under a vat of beer or wine, and toxic fusel oils vaporize first. Distillers call this the "head" of the distillation. It smells like nail polish remover. At the Plymouth gin distillery, they recycle it as an industrial cleaner. Next, as the temperature continues to climb, comes the "heart," the ethyl alcohol that is the goal of distillation. At the end of the run come the heavier molecules that contain additional toxins, but also some of the more flavorful compounds that make whiskey and brandy taste so good. This section, the "tail," must be cut off as well, but distillers may leave a little in to flavor their spirit.

Knowing where to cut the heads and tails is the mark of a good distiller. Homemade moonshine, bathtub gin, and other such amateurish attempts at distillation can be fatal because those dangerous compounds might not be extracted properly. Cheaper, mass-produced spirits may also produce worse hangovers if those toxins were not properly extracted or filtered out. Some liquors are double- or triple-distilled, meaning the heart is run back through the still to extract more heads or tails, and some, like vodka, are filtered through charcoal to remove the slightest impurity, leaving a clear and mostly odorless and tasteless spirit that is as close as possible to pure ethyl alcohol.

BUGS *in* BOOZE:
A SIX-LEGGED YEAST DELIVERY SYSTEM

Bugs in the brew? It is an age-old problem. Fermentation takes place in open tanks by necessity; otherwise, the pressure from the carbon dioxide would build to dangerous levels. But when a vat of fruit juice or grain mash is left to brew in an old barn or warehouse, bugs will surely find their way in. This is not always such a bad thing: lambic brewers in Brussels realize that some of their best strains of yeast come from insects falling from the rafters. In fact, yeast produce esters in order to attract insects, hoping they will pick up the yeast and move it around. This makes bugs unwitting accomplices in the dance between sugar and yeast.

HOW DID THEY GET THAT PEAR IN THE BOTTLE?

⊰ PEAR ⊱

Pyrus communis

ROSACEAE (ROSE FAMILY)

PEAR CIDER, OR PERRY, IS DELIGHTFUL WHEN YOU CAN GET IT.
The pears best suited to cider (called perry pears) tend to be small, bitter, dry, and more tannic than dessert pears. Pear cider is less common in part because pear trees are susceptible to a bacterial infection called fire blight, which is difficult to control; the disease has wiped out many old orchards. Pear trees also grow slowly and bear fruit later in life, making them a long-term investment rather than a quick crop, which is why farmers say, "Plant pears for your heirs."

Another issue is that once pears are picked they must be fermented immediately; they can't be stored like cider apples can. Pears also contain a nonfermentable sugar called sorbitol, which adds sweetness but has one drawback: for people with sensitive systems, it acts as a laxative. One popular English pear variety, Blakeney Red, is also called Lightning Pear for the way it shoots through the system. This quirk has earned cider pears yet another folk saying: "Perry goes down like velvet, round like thunder and out like lightning."

Having said that, real pear cider—as opposed to apple cider with pear flavoring added—is well worth seeking out. It is sweet but not cloyingly so, and it has none of the tartness and acidity that some apple ciders possess.

Pear brandy and *eau-de-vie de poire* are made in much the same way that apple brandy is, by distilling fermented pear mash or juice. Poire Williams is a popular French brandy made from Williams pears, which are known in the United States as Bartlett. It takes about thirty pounds of pears to create one bottle—and if that isn't labor-intensive enough, some pear brandies are sold with a pear inside the bottle. When the fruit is small, bottles are carefully slipped over them and hung from nearby branches for support, making the orchards especially difficult to tend as the pears ripen, inside glass, on the trees.

BARLEY

Hordeum vulgare
POACEAE (GRASS FAMILY)

Imagine a world without beer, whiskey, vodka or gin. Impossible! Yet it is no exaggeration to say that without barley, they wouldn't exist. Among grains, barley is uniquely well suited to fermentation, so much so that it can even help with the fermentation of other grains—making it possible to coax alcohol from the most unlikely of sources.

To understand the near-miraculous powers of barley, consider first the fact that cereal crops—barley, rye, wheat, rice, and so on—are not bursting with fermentable sugars the way apples or grapes are. Grains are packed with starch, which is a kind of storage system that allows plants to save the sugar they make during photosynthesis for some later use. To make alcohol from grains, the starch has to first be converted back to sugar.

Fortunately, it takes nothing but water to persuade a plant to perform this trick. Each individual grain is, after all, a seed. When that seed germinates, it's going to need some food to sustain it until it's big enough to put down roots, spread out leaves, and make its own dinner. That's what the stored sugar is for. All a brewer needs to do is to get the grain wet—a process called malting—which starts germination and prompts enzymes inside the grain to break down starch into sugar to feed the tiny seedling. Then it's simply a matter of adding yeast to devour the sugar and excrete alcohol. Simple, right? Not exactly.

Distillers learned the hard way that not every grain gives up its sugars so easily. That's where barley comes in: it possesses unusually high levels of the enzymes that convert starch to sugar. It can be mixed with another grain, like wheat or rice, to jump-start the process in those grains as well. For that reason, malted barley is the brewer's best friend—and has been for at least ten thousand years.

PART I

31

THE CLASSICS

the botany of beer

Barley is a type of tall, very tough grass that isn't bothered by cold, drought, or poor soil, making it widely adaptable around the world. In the wild, the spikelets of grains shatter and drop as soon as they are ready to germinate, but some enterprising early humans noticed that occasionally a barley plant would hold tightly to its grains. This was an ordinary genetic mutation that might not have had much benefit for the plant, but it was one that people liked: if the grains stayed on the stalk, they were easier to harvest.

And that is how the domestication of barley happened. People selected seeds that possessed a trait they liked, and those seeds went around the world. Barley originated in the Middle East, making its way to Spain by about 5000 BC and to China by 3000 BC. It became a staple grain in Europe. Columbus brought it to America on his second voyage, but it wasn't established in the New World until the late 1500s and early 1600s, when Spanish explorers took it to Latin America and English and Dutch settlers brought it with them to North America.

It is easy to imagine the ancient, happy accident that led to the invention of beer. Picture a bucket of barley left to soak overnight to soften the tough outer husk. Wild yeast would have found its way into the bucket, and someone would have thought to taste the strange, foamy mixture that resulted from the yeast going to work on all those sugars. There it was: beer! Yeasty, bubbly, mildly intoxicating beer. The priorities of people in the waning years of the Stone Age must have undergone a rapid reshuffling as society organized itself around the need to reproduce this glorious mishap on a larger scale. (Is it any wonder that the Bronze Age, with its large metal tanks, came next?)

By archeological standards, it took no time at all for sophisticated beer-making techniques to develop. Patrick McGovern, a University of Pennsylvania Museum archaeologist who studies the history of fermentation and distillation, analyzed the residue on pottery fragments found at the Godin Tepe site in western Iran. He detected the residue of barley beer on drinking vessels and was able to date it to 3400 BC to 3000 BC. He believes that the beer was probably not terribly different from what we drink today, except that it might

WHERE DO BEER AND WHISKEY GET THEIR COLORS?

WHISKEY DOES NOT NECESSARILY COME OUT OF THE BARREL WITH A DEEP AMBER COLOR, and beer is not always as dark in the fermentation tank as it is in the bottle. Caramel coloring is used in some beer and spirits to ensure the color is consistent from one batch to the next. Color is also used to indicate the age of a particular bottling: While an eight-year-old Scotch and a twenty-year-old Scotch may not come out of the barrel in different colors, the older Scotch can be tinted a darker color to suggest a longer aging period. In the case of beer, the color of the beer is closely associated with its branding: an amber is expected to be red, and a stout is supposed to be dark brown.

Purists consider caramel an unnecessary additive that should be dispensed with. So-called beer caramel, or Class III 150c caramel, is prepared with ammonium compounds and is one of two types of caramel that have come under criticism from consumer groups for the possibility that they contain carcinogens. (Class IV "soda caramel" is the other type made with ammonium compounds.)

Whiskey, on the other hand, is typically colored with "spirit caramel," or Class I 150a caramel, which is not made with ammonium compounds. Although it is not considered harmful and apparently does not change the flavor of the drink, some whiskey purists advocate for the return of "real whiskey" without unnecessary coloring. Highland Park Scotch boasts that it adds no coloring to its spirits; many smaller craft distilleries also avoid caramel color. In the United States, only blended whiskies are allowed to include caramel, but "straight whiskey" or "straight bourbon" may not.

not have been as finely filtered. Cave paintings and markings on the pottery depict people sitting around a large pot of beer and drinking through long straws. The straws were aimed at the middle of the brew so that whatever sediment sank to the bottom or rose to the top could be avoided.

Beer making grew more sophisticated in Roman times. Roman historian Tacitus, describing German tribes, wrote that "for drink they use a liquid made of barley or wheat and, by fermentation, given a certain likeness to wine." It wasn't long after that, perhaps as early as 600 AD, that people in barley-growing regions realized that just like wine and cider, beer could be distilled into a much more potent spirit. By the late 1400s, whiskey—which was then called aqua vitae, a generic term for distilled spirits—was being made in the British Isles.

growing the perfect barley

Although the debate between the Irish and the Scottish over who invented it may never be settled, the fact is that whiskey was born in that region precisely because its climate and soil is so well-suited to growing barley. Stuart Swanston, a barley researcher with the Scottish Crops Research Institute, believes that Scotland's chilly weather is perfect for its most famous crop. "The advantage we have on the eastern coastal strip of Scotland is that we're close to the North Sea," he said. "We have mild winters and rubbishy summers—it's a long, cool, humid growing season. That means lots of starch in the grain, which is very suitable for high levels of alcohol." But if the weather is off and the starch doesn't form perfectly in the grain, it gets fed to animals and Scotland's finest distillers have to turn to France or Denmark for the quality of grain they require.

The type of barley best suited to brewing and distilling is a matter of some debate. Barley is classified as two-row or six-row: two-row barley has one row of grain on either side of the seed head, and six-row barley has three rows on each side. The six-row barley is the result of a genetic mutation that proved popular in Neolithic times because it yielded more grain per acre and contained more protein. Two-row barley, on the other hand, contains less protein but more starch to convert to sugar. This makes it less suitable as food but perfect for brewing and distilling. Although it is traditional for European brewers and distillers to use two-row barley, many

WHISKEY OR WHISKY? OR, HOW TO DRIVE A COCKTAIL WRITER MAD

THE WORD ORIGINALLY CAME FROM THE GAELIC *uisgebeatha*, meaning "water of life." That became something like *whiskybae*, and shortened versions like *whiskie* and the even more cheerful *whiskee* were in use in the early eighteenth century. By the nineteenth century, *whisky* was the Scottish and British spelling, whereas *whiskey* was the favored spelling in Ireland and the United States. (However, U.S. liquor regulations go with *whisky* in all but one instance.) *Whisky* is also used in Canada, Japan, and India.

Some writers go to great lengths to switch back and forth, even in the same sentence, depending on whose hooch they are referring to. Others stick to the spelling in use in their own country on the grounds that an American would not use *colour* to refer the hue of a British carpet or *aubergine* to describe an eggplant enjoyed in London. In these pages, "whisky" is employed only to refer specifically to a product distilled in one of the countries that shun the e.

Americans nonetheless favor the six-row variety, in part because it's more widely available. Six-row barley also tolerates a wide range of climates across the country, making it easier to grow on a large scale.

Barley is further categorized into spring and winter types based on growing season. A winter barley can be planted in fall and harvested in spring, while a spring variety will be planted in spring and harvested in summer. The spring type is traditional among brewers, but modern genetics has demonstrated that there may, in fact, be little difference between the two.

What does matter is the weather and the soil. Even the type of fertilizer used in the fields can make a difference: too much nitrogen in the soil translates to too much nitrogen in the grain, which

GROW *your* OWN

BARLEY

Even the most dedicated brewers would probably not bother growing their own barley, but it can be done. A hundred-square-foot plot will produce about ten pounds of barley, enough for a respectable five-gallon batch of home-brewed beer.

The best way to prepare a small garden patch for grain growing is to start in the fall. Clear the area of weeds, but don't dig into it. Instead, cover the ground with several overlapping layers of cardboard or newspaper (use entire sections of newspaper so layers are at least twenty pages thick); water well to make the paper stay in place; and cover with layers of manure, compost, grass clippings, dried leaves, rice straw, or bagged soil mixes. Make your pile a foot tall or more. It will settle considerably over the winter.

In spring, clear the bed of any weeds that might have appeared and cover with a thin layer of compost. On a day when the ground is dry, scatter the seeds and simply rake them in lightly and water. (You'll need about three-quarters of a pound of seeds.) Water until late summer, and then let the plants turn a golden brown.

When the grains are hard and dry, cut them down and gather the stalks into bundles. Once they are completely dry, they can be threshed by laying them out on a clean surface and whacking them with any blunt wooden instrument. (A broom handle works.) The traditional way to clean the grain, called winnowing, is to go outside on a windy day and pour the grain from one bucket to another, allowing the dried straw to blow away.

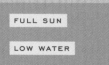

FULL SUN

LOW WATER

HARDY TO -10F/-23c

increases the level of protein and pushes down the amount of starch. "Too much protein can be a bad thing for brewing traditional ales and whisky," Swanston said. "But if you're simply making malted barley that you're going to add to other grains, more protein is actually better. It has even more of the enzymes that help break down starch in other grains."

on malting

Another important natural resource contributes to the extraordinary character of Scotch whisky in particular: peat. The bogs form as a result of thousands of years of slowly decaying plant debris. Peat logs, sliced neatly out of the bogs, have been used for centuries as a slow-burning source of fuel—and have played a key role in the malting of barley for distillation.

Traditionally, wet barley grains would be spread on the floor of a malting house and allowed to sprout for about four days, during which time the enzymes in grains gobble up oxygen to help them break down the sugars, and release some of the stored carbon in sugars as carbon dioxide. They naturally heat up during this process, so workers rake through them and turn them over to keep them cool and to prevent the young roots from becoming entangled. At this stage, the barley is called green malt.

After the barley has been dampened and allowed to sprout, it must be heated to stop germination, essentially killing the young seedling while capturing the newly released sugar. A fire made of peat logs gently dries the grains over a period of about eight hours, and the smoke infuses them with that delightfully dark, earthy flavor that good Scotch is known for. At least, that's how it used to work. Only a handful of distilleries, including Laphroaig, Springbank, and Kilchoman, still malt and peat their own barley, a practice known as "traditional floor malting." Today most Scottish distilleries order their barley from large, commercial malting houses that pipe peat smoke through the grains at whatever level the distillery requests. This allows less peat to be used, helping to conserve bogs. Whiskey makers around the world order peat-smoked barley from Scotland if they want to achieve that distinctive flavor.

RUSTY NAIL

Drambuie is a rich and gorgeous liqueur made of Scotch, honey, saffron, nutmeg, and other mysterious spices. Like many such concoctions, it is unnecessarily burdened by one of those legends only a marketing executive could love: In 1745, Charles Edward Stuart, otherwise known as Bonnie Prince Charlie, tried to regain the throne after his father's ouster by opponents. He was given refuge on the Isle of Skye and, according to the story, gave his cherished drink recipe to his protectors as a show of thanks. It changed hands a few times before becoming the commercial product it is today.

Dethroned princes aside, Drambuie is excellent on its own, over ice as an after-dinner drink, and as an ingredient in one of the world's simplest and most enjoyable cocktails. A Rusty Nail is the perfect gateway drink for anyone not quite ready for the bracing woodsiness of Scotch. (Not to be outdone, the Irish have their own whiskey liqueur. The Irish Mist backstory is even mistier than Drambuie's, involving an ancient manuscript brought to Ireland by a mysterious traveler and then handed down through several generations. It is a similarly sweet and spicy liqueur, and although it is not as popular as Drambuie, devotees of Irish whiskey should give it a try.)

Because this recipe combines Scotch with a Scotch-based liqueur, it also illustrates a clever bartending technique: whenever possible, mix spirits with liqueurs made from the same base spirit.

1 ounce **Drambuie**
1 ounce **Scotch**

Add the ingredients to an Old-Fashioned glass half filled with ice and stir. The Irish version, a Black Nail, is made with Irish Mist and Irish whiskey.

Once the barley is malted and dried, it is usually allowed to rest for a month or so before being mixed with water and yeast to make a mash. It ferments for a couple of days, and then the beerlike liquid—a wash—is separated from the spent grain. It goes into the still at about 8 percent alcohol, and from there it is distilled into whisky.

breeding a better grain

Botanists around the world are working on developing new varieties of barley that are even better suited for making beer, whiskey, or malt extract. The Scottish Crop Research Institute is tackling the problem of mildew diseases, such as fusarium, the same ailment that causes black spots on roses. Because European farmers in particular are restricted on the types of chemicals they can spray on their crops, a mildew-resistant barley would be enormously useful. And at the University of Minnesota, botanists are also tackling the fusarium problem and introducing new varieties to American brewers, who already use the university's barley strains in about two-thirds of all beer produced in the country.

Today's breeding programs are nothing more than a continuation of the last ten thousand years of human intervention. Stuart Swanston said, "Barley is grown from northern Scandinavia to the foothills of the Himalayas, from Canada to the Andes. It's made this phenomenal journey from the Fertile Crescent and spread throughout the world with amazing adaptability."

THE MAGIC OF SCOTCH AND WATER AND THE CONTROVERSY OVER CHILL FILTRATION

THE BEST WAY TO DRINK WHISKEY—and any other high-proof spirit, for that matter—is with a little splash of water. Scotch connoisseurs recommend adding five or six drops per ounce. It doesn't dilute the flavor; it actually heightens it.

To understand why, consider the fact that the molecules with the most flavor—larger fatty acid molecules that come through near the end of the distillation—tend to break away from the alcohol in the presence of water and form a suspension. So a splash of water will cause some whiskey to become cloudy—and those clumps of molecules in suspension bring the richest flavors forward. (Dribbling ice water into absinthe causes cloudiness for much the same reason, but more on that later.)

Even storing whiskey at low temperatures can cloud it. Whiskey is generally not sold at cask strength; it comes out of the barrel at a higher proof and is watered down to, say, 40 percent alcohol by volume before bottling. Once that water is added, the fatty acid molecules are even more likely to break loose under colder temperatures and form a cloudy suspension in the bottle that distillers call chill haze.

To get around this, many whiskey makers put their spirit through a chill filtration process in which the temperature is deliberately lowered to force those fatty acids to clump together so that they can be sifted out with a metal filter. While this does prevent cloudiness, some whiskey lovers believe that chill filtration, like caramel coloring, is another unnecessary artifice that interferes with flavor and should be done away with. Ardbeg, an Islay Scotch, states plainly on its label that the product is not chill-filtered, and Booker's Bourbon also brags that it is unfiltered.

Next time you're sitting in a bar, show off your chemistry prowess by adding water to your whiskey to check for the presence of long-chain fatty acid molecules—then raise the glass and enjoy.

BUGS *in* BOOZE: EARTHWORMS

Connoisseurs of Scotch will run across a strange term from time to time in whisky reviews. A particularly pungent, heavy, malty spirit might be described as having a distinctive worm flavor. Given the fact that earthy peat smoke is such a predominant flavor in Scotch whisky, it is not too much of a stretch to imagine that a few earthworms got into the mix as well.

But to distillers, a worm is a coiled copper tube submerged in water. This particular condensing technique is just another way to subtly alter the flavor of the spirit through the shape of the still and the manner in which flavors are extracted. Some distillers claim that the use of a "worm tub" does give a meatier flavor to the finished product—but no actual worms are harmed in the making of whisky.

Which is not to say that worms have never been used in boozy, medicinal tonics. This 1850s era recipe for the treatment of "Eyaws" (presumably yaws, a nasty bacterial infection of the skin and joints) from the archives of Kentucky farmer John B. Clark calls not just for earthworms but for other frightful ingredients. If it didn't cure people, it would certainly knock them back into bed for a few more days.

Recipte for the Eyaws

Take 1 pint of hogs Lard
1 handfull of earth worms
1 handfull of Tobacco
4 pods of Red pepper
1 spunfull of Black pepper
1 Race of Ginger

Stew them well together, & when Applyed mix Sum Sperits of Brandy with it.

CORN

Zea mays

POACEAE (GRASS FAMILY)

Very little good news came out of the Jamestown colony in the early days. The settlers suffered starvation, disease, drought, and horrific accidents. Crops failed and supplies were slow to arrive from England. It must have been nice, then, for John Smith, one of the organizers of the effort to establish a settlement, to get a letter in 1620 from colonist George Thorpe that included this cheerful line: "Wee have found a waie to make soe good a drinke of Indian corne as I protest I have for divers times refused to drinke good stronge Englishe beare and chosen to drinke that." Apparently, there was just enough copper among their meager supplies to build a still. Corn whiskey was one of the first innovations to come from the struggling Virginia colony.

Corn—called maize by Columbus, who might have heard the word *mahis* from the Taíno people in the Caribbean—was a revelation to the Europeans. (At the time, the word *corn* referred to any sort of grain, so Europeans called it Indian corn to distinguish it from wheat, millet, rye, barley, and other grains.) Columbus brought it back from his voyages and it quickly went into cultivation in Europe, Africa, and Asia. It was easy to grow, adaptable, and—best of all— the grains could be saved for winter. As Thorpe learned, it made a nice drink, too.

chicha and cornstalk wine

In Mexico, archeological evidence points to corn as a dietary staple as early as 8000 BC. Its range extended into parts of Central and South America, where every culture found different uses for the plant. When the Spanish arrived, two fermented beverages were widespread: corn beer, made from the ripe yellow kernels, and cornstalk wine, made from the sweet juice of the stalk. Exactly when these traditions began, and what sort of wild *Zea* might have been used, are questions that continue to vex archeologists.

Corn was domesticated so long ago that its ancestor no longer survives. Botanists assume that early corncobs were much smaller, the size of a finger, perhaps. They probably resembled their cousins in the *Zea* genus, many of which look like ordinary tall grass with an unremarkable seed head. These weedy relations are called teosinte. They look nothing like modern corn. Instead of producing a sturdy central stalk, they take the form of a wide, bushy clump of grass. The seed heads hold five to ten small seeds in a straight line, as opposed to a few hundred arranged around a corncob.

A team of archeologists led by Michael Blake at the University of British Columbia now believe that early corn might have been selected and domesticated not for its grain but for its juice. Cornstalk quids—bits of plant fiber that were chewed and then spit out—have been found at archeological sites dating to 5000 BC, suggesting that people prized the plant for its sweetness. And analysis of human remains found at those sites indicates that they were getting corn sugar in their diet but not much corn grain.

Over time, through some combination of human selection, chance hybridization, and mutation, corn came to resemble the plant we know today. When Columbus saw it for the first time, the ears might have been smaller, but it would have been obvious that its real value came from the kernels of corn, not sugar from the stalk. Columbus brought a new sweetener to the Americas in the form of sugarcane, and from that time on, cornstalk sugar declined in importance.

But cornstalk wine didn't disappear entirely: A few centuries later, Benjamin Franklin wrote that "the stalks, pressed like sugar-cane, yield a sweet juice, which, being fermented and distilled, yields an excellent spirit," suggesting that the practice was still alive. Even today some tribes, such as the Tarahumara of northwest Mexico, continue to make the wine as a traditional tribal practice. The stalks are pounded against rocks to extract the juice, which is mixed with water and other plants, then naturally fermented and consumed within a few days.

Corn beer, called chicha, was the other corn beverage Europeans encountered. Its exact origins are a bit of a mystery, but the rather sophisticated process was already centuries old when the Spaniards arrived, and the tradition continues today. Like other grains, the starch in corn has to be converted to fermentable sugar before the yeast can go to work on it. In Peru and surrounding areas, it is made by chewing uncooked, ground corn, then spitting it out and mixing the wads of chewed corn with water. Digestive enzymes in saliva are effective at converting starch to sugar, so the spit was an integral part of the process.

Archeologist Patrick McGovern, who studies the ancient origins of alcoholic beverages, worked with Dogfish Head brewery in Delaware to brew a batch using the traditional method. The experiment reads like the setup for an old joke: two anthropologists, a brewer, and a reporter from the *New York Times* walk into a bar. But what happened next was no joke. Behind the bar was a batch of ground purple Peruvian corn that they planned to chew, spit out, and mix

THE SEX LIFE *of* CORN

Next time you pull a piece of silk from between your teeth while you're eating a fresh ear of corn, remember that you've just spat out a fallopian tube. Corn has a curious anatomy: the tassel at the top of the plant is the male flower; when mature, it produces two million to five million grains of pollen. The wind picks up those grains and moves them around.

The ear of corn is actually a cluster of female flowers. A young ear contains about a thousand ovules, each of which could become a kernel. Those ovules produce "silks" that run to the tip of the ear. If one of them catches a grain of pollen, the pollen will germinate and produce a tube that runs down the silk to the kernel. There the egg and pollen grain will meet at last. Once fertilized, that egg will swell into a plump kernel, which represents the next generation—or a bottle of bourbon, depending on your perspective.

with a traditional recipe of barley, yellow corn, and strawberries. But chewing the corn was almost unbearable: the reporter compared the texture to uncooked oatmeal, and the wads they spat out "resembled something a cat owner might be familiar with, if kitty litter came in purple." One very small batch came out of the experiment, and that was the end of it. With a brewery full of modern equipment, chewing raw corn was clearly not worth the effort.

Apparently Dogfish Head is not the only brewery to come to this conclusion. The chicha sold today in Latin America is made using methods more similar to modern beer making. Like pulque, the agave-based beer, chicha is made on a small scale, served fresh, and often flavored with fruit and other sweeteners.

the birth of bourbon

It was a small step from corn beer to corn whiskey. Early settlers realized that corn was the easiest grain to grow in the unfamiliar terrain they found themselves in. Fortunately, they had the example of experienced local Indian farmers to follow. Clearing fields with nothing but hand-forged tools was backbreaking work, so they must have been relieved to learn that corn could be sown among tree stumps. Transporting a crop to market was difficult as well; farmers had to find a use for their corn close to home. An early corn beer, perhaps brewed with molasses imported from the Caribbean, was one popular solution. From that a fairly rough, crude whiskey was born—and bourbon was not far behind.

Because it was such an easy crop, planting a cornfield became the standard method for settlers to establish a claim on their land. Some early land grants, particularly in Kentucky, were contingent upon the settlers building a permanent structure or growing corn. This bit of history has led to a sort of creation myth about Kentucky and bourbon. Distillers and bourbon enthusiasts are fond of claiming that Thomas Jefferson, as governor of Virginia, offered sixty acres to anyone who would plant corn. In fact, the Virginia Land Law, passed the month before Jefferson took office, offered four hundred acres to settlers who could prove their claim, and planting a crop of corn was just one of several ways they could demonstrate that they'd settled the land. Nonetheless, the idea that a Founding Father practically ordered Kentucky to become the glorious land of bourbon that it is today makes for a better story.

TYPES OF CORN

DENT (*ZEA MAYS* VAR. *INDENATA*):
A cross between flint and flour, this is a softer corn with a dent on either side of the kernel. Also called field corn, dent is the most widely grown type in the United States.

FLINT (*ZEA MAYS* VAR. *INDURATA*):
With a hard outer layer and soft endosperm, this corn has a lower yield but matures earlier than other varieties.

FLOUR (*ZEA MAYS* VAR. *AMYLACEA*):
A soft corn primarily ground into flour.

POD (*ZEA MAYS* VAR. *TUNICATA*):
An old Peruvian variety in which each kernel is protected by its own husk.

POP (*ZEA MAYS* VAR. *EVERTA*):
This corn has a large endosperm that, when heated, explodes and turns the grain inside out so that the translucent shell is on the inside.

SWEET
(*ZEA MAYS* VAR. *SACCHARATA* OR *ZEA MAYS* VAR. *RUGOSA*):
A soft, high-sugar corn grown for canning or eating fresh.

WAXY (*ZEA MAYS* VAR. *CERATINA*):
A variety discovered in China in 1908 that contains a different kind of starch. It is used as an adhesive and in food processing as a thickener and stabilizer.

Kentucky had a few other things going for it besides an abundance of corn. Many of the state's early immigrants came from Scotland and Ireland, so they knew their way around a still. (That information exchange went both ways. By the 1860s, corn had become the grain of choice for Scottish distilleries as well.) The state had another natural resource that lent itself to whiskey making: rich limestone deposits from which clear, cool spring water flowed. Settlers were more likely to set up camp near a spring, so it's no surprise that early distilleries were built there, too. Among the many benefits of "limestone water" was the fact that it came out of the ground at about 50 degrees Fahrenheit, the perfect temperature for the cooling and condensation process in the days before refrigeration. The higher pH level inhibited iron particles that can give whiskey a bitter taste. And it is possible that the elevated levels of calcium, magnesium, and phosphate encouraged the growth of lactobacillus, a bacteria that plays a role in fermentation. Although corn continued to be made into a rough moonshine around the country, Kentucky took advantage of its natural resources and started building a respectable whiskey industry.

Kentucky produces about 90 percent of the world's bourbon supply. A recent surge in its popularity has created a booming export market, with distilleries working at full capacity and a tourist-friendly Kentucky Bourbon Trail attracting visitors to the state. The water is still one of the state's claims to fame, but not all bourbon is made with natural spring water today—larger facilities use filtered river water instead. University of Kentucky hydrogeologist Alan Fryar has analyzed the role of water in bourbon; he believes that there is some scientific basis for claiming that the limestone water is superior, particularly when it comes to inhibiting iron in the water, but much of its value cannot be quantified. "It gets to the idea of *terroir*," he said. "Our water is used to grow the corn, it's used for cooling, it's used in the mash. How exactly that changes the flavor is almost impossible to quantify—but it is important." Distillers will always make much of Kentucky's good water: bourbon industry expert James O'Rear was once quoted as saying that "limestone in bourbon lets you wake up the next morning feeling like a gentleman."

OLD-FASHIONED

1½ ounces bourbon
1 sugar cube
2 to 3 dashes Angostura or orange bitters
Maraschino cherry or orange peel (optional)

Place the sugar cube in the bottom of an Old-Fashioned glass and give it a few dashes of bitters. Add a splash of water and use a muddler to crush the ingredients together. Swirl the mixture around in the glass, add bourbon and ice, and stir. Although the addition of fruit to this drink is considered a sacrilege in some circles, a true Italian maraschino cherry perfectly complements bourbon's natural sweetness.

HAVE A NICE GLASS OF CORN

BLENDED WHISKEY: Although definitions vary around the world, blended whiskies may contain some corn. Suntory's Hibiki and Royal brands, for instance, include corn and other grains.

BOURBON: An American-made corn-based whiskey aged in new charred oak barrels. Must contain at least 51 percent corn. *Straight bourbon* is aged for at least two years, with no added color, flavor, or other spirits. *Blended bourbon* must contain at least 51 percent straight bourbon, but may also contain added color, flavor, or other spirits.

CHICHA DE JORA: A South American fermented corn beer. *Chicha morada* is a nonalcoholic version.

CORN BEER: Some beers contain corn as an adjunct ingredient, making up 10 to 20 percent of the mash. Examples of beers containing corn include China's Harbin Beer, Mexico's Corona Extra, and Kentucky Common Beer, a style that includes about 25 percent corn, still made by specialty brewers today.

CORN VODKA: Craft distillers are making excellent corn-based vodkas. Tito's Handmade Vodka, from Austin, Texas, is a fine example.

CORN WHISKEY: Similar to bourbon but must be at least 80 percent corn. Can be unaged or aged in used or uncharred new oak containers.

MOONSHINE OR WHITE DOG: A catchall term for unaged whiskies, which were made of corn historically and often still are.

PACIKI: A Mexican cornstalk beer.

QUEBRANTAHUESOS: The name means "bone breaker." A Mexican drink of fermented cornstalk juice, toasted corn, and the seeds of the Peruvian pepper tree (*Schinus molle*).

TEJATE: A nonalcoholic brew of corn, cacao, and a few other ingredients, made in and around Oaxaca.

TEJUINO: A fermented (and only very mildly alcoholic) Mexican cold drink made of corn dough, widely sold today.

TESGÜINO: A traditional corn beer from northern Mexico.

TISWIN: A southwestern pueblo beer made of corn, sometimes combined with cactus fruit, roasted agave juice, or other ingredients.

UMQOMBOTHI: A South African beer made of corn and sorghum.

choosing the perfect corn

While wines are known primarily by the varieties of grapes used, distilleries have not, until recently, explored unique strains of heritage corn. The grain is still seen as a commodity; whiskey is typically made with No. 1 or 2 Yellow Field Corn, a standard designation that measures only the color and "soundness" of the grain—the amount of grain in a bushel that is undamaged, uninfested, and free of debris. But why not use artisanal, heirloom corn varieties to make bourbon? Master distiller Chris Morris, the brains behind Woodford Reserve's line of extraordinary, award-winning bourbons, said, "We just want big, clean, dry corn. The starch is what it's all about. The corn is pretty much just the muscle that we use to make alcohol. We've done distillations with various types of corn, and basically, corn's corn. We even experimented with organic grain and we just couldn't tell a difference."

But Joel Elder, from Tuthilltown Spirits in Gardiner, New York, sees it another way. "People say that the artisanship is in the distilling, and I fundamentally disagree. My opinion is that the distillation is the easiest part. The further back you go in the process—fermentation, grain handling and storage, growing—the more artisanal it becomes. Look at wine. In wine, we talk almost exclusively about the grape. Nobody does that with bourbon." He has experimented with a number of different heirloom corns, including Wapsie Valley, known for producing red kernels. (The legend behind Wapsie Valley is that at corn shuckings, any man who found a red kernel could kiss the girl of his choice, and Wapsie Valley could turn an innocent gathering into a free-for-all.) He also grows Minnesota 13, a dent corn that was widely used for moonshine during Prohibition. "We get a big, buttery popcorn flavor out of these," he said. "Does the variety of corn make a difference? I can run separate distillations of just those two varieties and make a believer out of anyone."

THE CORK OAK

THE NATIVE PORTUGUESE OAK *Q. SUBER* provides another essential ingredient to wine and spirits: the cork. The trees live for more than two hundred years, and by about age forty, they have produced enough of their thick, spongy bark to produce a harvest of four thousand corks. Stripping the trees of their bark does not hurt them because the bark regrows. In fact, cork growers argue, the ability to harvest the bark provides an economic incentive to leave massive stands of old oak trees intact.

The increased use of screw tops and synthetic caps has hurt the cork industry in Portugal, Spain, and North Africa, where most cork forests are found. Growers insist that the natural corks are not only more authentic and better for the wine but are actually kinder to the environment than a synthetic replacement.

ANGELS' SHARE: IN STORAGE, A SMALL AMOUNT OF ALCOHOL ESCAPES THE BARREL THROUGH EVAPORATION. DISTILLERS CALL THIS LOST ALCOHOL THE ANGELS' SHARE. WHISKEY AND BRANDY MAKERS ESTIMATE THAT THE ANGELS GET ABOUT 2 PERCENT OF THE ALCOHOL IN A BARREL EACH YEAR, ALTHOUGH THAT CAN VARY DEPENDING ON HUMIDITY AND TEMPERATURE. FORTUNATELY, THEY CAN AFFORD TO LOSE SOME, AS MOST SPIRITS ARE AGED AT A HIGHER PROOF THAN THE FINAL BOTTLING. (SOME WATER IS LOST AS WELL, WHICH CAN HELP KEEP THE OVERALL PERCENTAGE OF ALCOHOL FROM DROPPING TOO MUCH.)

ONE RESULT OF THIS SLOW LEAKAGE OF ALCOHOL IS THAT IT ATTRACTS A STRANGE CREATURE RARELY SEEN OUTSIDE OF DISTILLERIES. THE BLACK FUNGUS *BAUDOINIA COMPNIACENSIS* FEEDS ON ETHANOL AND APPEARS AS A BLACK STAIN ON THE WALLS OF CAVES AND WAREHOUSES WHERE SCOTCH AND COGNAC ARE STORED. EUROPEAN DISTILLERS IN PARTICULAR ARE NOT BOTHERED BY IT; IN FACT, IT IS SEEN AS A FRIENDLY COMPANION AND A MARK OF AUTHENTICITY.

OAK

Quercus spp.

FAGACEAE (BEECH FAMILY)

NOTHING TAMES A ROUGH SPIRIT LIKE AN OAK TREE. The practice of aging whiskey or wine in a barrel might have started as a practical solution to a storage problem, but it was soon obvious that something wonderful happens when alcohol comes in contact with wood—and oak in particular.

Oak trees have been around for about sixty million years. They emerged as a distinct genus not long after the mass extinction of dinosaurs. Taxonomists disagree about the exact number of species; depending you who you ask, the number ranges from sixty-seven to six hundred. However, we are only concerned with the handful of American, European, and Japanese species used by barrel makers for wine and spirits.

Wooden barrels have been in use for at least four thousand years, judging from archeological evidence, and oak was probably the natural choice from the beginning. The wood is hard, dense, but still pliable enough to bend into a slight curve. It was used for shipbuilding, and surely one of the first pieces of cargo to be loaded on the ship was a barrel of wine for the crew.

What the first barrel maker might not have known was that the anatomy of oak is perfectly designed not just to hold but also to flavor the liquid it contains. Oak trees are "ring porous," which means that the vessels that carry water up the tree are found in the outer growth ring. As the tree matures, the older vessels become plugged with crystalline structures called tyloses, and as a result, the center of the tree—the heartwood—doesn't conduct water at all, making it well suited for use as a watertight barrel. American oaks are particularly rich in tylose as compared to European oaks. In fact, the European trees have to be carefully split along the grain, rather than cut, in order to avoid rupturing vessels and creating a leaky barrel.

The trees also happen to produce an astonishing array of flavor compounds that break free from the wood in the presence of alcohol. European oak, *Quercus robur* in particular, is high in tannins, which give wine a certain roundness and full-bodied quality. American white oak, on the other hand, releases the same flavor molecules found in vanilla, coconut, peach, apricot, and cloves. (In fact, artificial vanilla is made from a sawdust derivative because it has such high levels of vanillin.) Those sweet flavors might not be what a winemaker is looking for, but they are pure magic in bourbon.

Perhaps the most important influence on oak-aged spirits comes not from the tree but from the barrel makers, called coopers. They learned that coaxing oak staves into gentle curves required two things: time and heat. Freshly cut oak is given time to dry, which not only makes it easier to work with but also concentrates those important flavors. The staves are also lightly cooked to make them more pliable as they are shaped, and fire causes some of those flavors to caramelize, so that caramel, butterscotch, almond, toast, and warm, woodsy, smoke essences emerge.

Some whiskey barrels are entirely burned inside. No one knows how this started. It's possible that a cooper accidentally lit a bigger fire than intended and decided to use the barrel anyway. Perhaps thrifty distillers burned the inside of old barrels used to store salted fish or meat to eliminate the flavor before filling it with whiskey. Regardless, the layer of charcoal filters and flavors the whiskey, particularly as the wood expands and contracts with changes in the weather. The Lincoln County process, made popular by Jack Daniel's, takes this one step further by burning sugar-maple wood and filtering the whiskey through ten feet of charcoal before it ever reaches the barrel.

The coopers made one more contribution: after Prohibition, when it became necessary to enact new laws regulating the now-legal liquor industry, they helped ensure that bourbon (and other whiskey) would, as of July 1, 1936, have to be stored in charred new oak containers in order to claim the name. This, the newly formed Federal Alcohol Administration claimed, distinguished "American-style whiskey" from Canadian products, which possessed a milder flavor owing to the fact that they were distilled at a higher proof and stored in reused cooperage. Although the law has gone through some revisions and challenges, the requirement to use new barrels for each batch of bourbon has been in place continuously, with only a brief respite from 1941 to 1945 because of wartime shortages.

One result of this quirk in American law is that there are an abundance of used bourbon barrels for sale. Scotch distillers love them: they employ a blend of used bourbon, port, and sherry barrels to give a nice complexity to their fine spirits. In fact, the Laphroaig distillery boasts that it uses exclusively Maker's Mark barrels. Used bourbon casks are also used to age rum and other whiskey blends.

The particular way that oak absorbs and releases a spirit has led to a great deal of experimentation. Coopers can construct barrels with trees grown in a particular climate or soil type, which can affect the tightness of the grain and the levels of tannins and flavor molecules. They can even build a barrel of sapwood as opposed to the denser, less absorbent heartwood. Distillers are starting to market whiskey aged in a barrel from one part of the tree or another, knowing that connoisseurs will literally lap it up.

A FIELD GUIDE *to* OAK

Q. ALBA: American white oak; grown in eastern United States, used for whiskey and wine.

Q. GARRYANA: Oregon oak, used by some Pacific Northwest wineries and distillers. More comparable to French oak.

Q. MONGOLICA: Japanese oak, popular among Japanese distillers.

Q. PETRAEA: Sessile or French oak; grown in Vosges and Allier. Preferred by winemakers.

Q. PYRENAICA: Portuguese oak, often used for port, Madeira, and sherry.

Q. ROBUR: European oak; grown in Limosin. Preferred for Cognac and Armagnac.

BUGS *in* BOOZE: ALKERMES SCALE

--- *Kermes vermilio* ---

Scale is a tiny insect that latches onto a branch and hides under its protective shell. This particular species preys on *Q. coccifera,* a species of oak tree in the Mediterranean. The females suck down tree sap until they are as big and round as a tick, secreting a crimson gummy exudate. At some point, in the course of scraping scale off the tree, somebody must have noticed that the red pigment stained clothing and hands. It was certainly well known to the Greek physician Dioscorides, who wrote an odd entry in *De Materia Medica* (50–70 AD) about little insects that grow on oaks and are "similar in shape to a little snail, which the women there gather by mouth." Dioscorides was known to get a few things wrong; it's unlikely that women actually picked bugs off trees by mouth when a stick would do. Even using a stick was tricky: the bugs had to be removed more or less intact, then killed (usually by steaming them or dropping them in vinegar), and then dried and taken to market, where they would be sold as a fabric dye.

Like most strange and unusual things in the natural world, the red pigment found its way into Italian liqueurs. The recipe can be traced back to an eighth-century medicinal tonic called *confectio alchermes* that called for taking a length of silk that had been dyed red with the insect, soaking it in apple juice and rose water to extract the dye, and adding extraordinarily rare spices that included ambergris (sperm whale bile), gold flakes, crushed pearls, aloe, and cinnamon. Over time, the recipe changed to include more familiar flavors like cloves, nutmeg, vanilla, and citrus, and the red dye came from cochineal, another insect-based dye newly arrived from the Americas. It was brighter in color and easier to harvest.

By the nineteenth century, the bright red liqueur known as alkermes (or alchermes) was made by several Italian distillers as a digestif, not a medicine. It also became the flavoring for a layered sponge-cake desert called *zuppa inglese*. A modern version of the liqueur is still available in Italy and can be found in specialty Italian food shops. The ancient Santa Maria Novella pharmacy in Florence carries their own formulation. Unfortunately, alkermes made with actual kermes scale are a thing of the past; the only red insect-based food dye allowed in the European Union is E120, the cochineal scale.

GRAPES

Vitis vinifera

VITACEAE (GRAPE FAMILY)

Quick: name a fruit that is made into alcohol. What comes to mind first? Probably grapes. But believe it or not, the very existence of grapes is surprisingly unlikely. The fossil record shows that grapes were established in Asia, Europe, and the Americas fifty million years ago. But when the last ice age, the Pleistocene epoch, began about 2.5 million years ago, vast sheets of ice covered much of the grape's range and nearly drove it to extinction. The vines that managed to pass the time in unfrozen corners of the world were the only ones left for early humans to encounter. It's entirely possible that the grapes that flourished before the ice age were far more diverse and interesting than what we grow today.

To make the success of the grape even more improbable, those early vines would have yielded nothing like the abundant clusters of sweet, marble-sized fruit we know today. The grapevines that survived the ice age were dioecious, meaning that each plant was either male or female. The vines depended upon insects to transport their pollen, and if a female was too far away from a male, it simply wouldn't happen. The fruit from these couplings was unpredictable as well. Grapevines, like apples, can produce offspring whose fruit will be quite unlike that of its parents. Some of those grapes would have been small, bitter, and full of unpalatable seeds.

So what happened to improve the grape's prospects? A mutation that changed the plant's sexual orientation. In dioecious plants, the females are female because a gene suppresses the formation of male anatomy and vice versa. But sometimes those genes go awry and nature creates a hermaphrodite. The vines that resulted from those mutations had both male and female anatomy on the same plant. Because the pollen didn't have as far to travel, the vines produced more abundant fruit. The earliest agrarians might not have understood why certain vines were more prolific, but they would have selected

NOBLE ROT

THE FUNGUS *BOTRYTIS CINEREA* INFLICTS GRAPES WITH A NASTY DISEASE called botrytis bunch rot. If it attacks in early spring, it can make leaves wither and flowers drop off the vine. On young, immature fruit it forms nasty brown lesions that turn black and split the fruit apart. The rotten grapes, filled with fungus, drop to the ground and wait for an opportunity to reinfect the plant. Botanists refer to the dead, infested fruit as mummies.

But sometimes, under just the right weather conditions, botrytis hits late in the season and causes something remarkable to happen. If temperatures stay between 68 and 78 degrees Fahrenheit, and if the humidity is very high, and if the grapes are just ripe enough, they are infected but not ruined by the fungus. Then, in order for the magic to happen, the humidity must drop to about 60 percent. In other words, it needs to be cool, and rain, and then stop raining, as the grapes are getting ripe.

If all this happens at just the right time, the fungus will dehydrate the grape, concentrate the sugars, but not destroy it. That's called noble rot, and it's responsible for some of the great botrytized wines of the world. Sauternes, made in a particular region in Bordeaux with Sémillon, Sauvignon Blanc, and muscadelle grapes, are the finest expression of what noble rot can do to a grape. They are sweet but faintly spicy, with a distinctive honey and raisin flavor. The wines can be expensive: noble rot is unpredictable, each grape must be picked individually by hand, and an entire vine might only yield the equivalent of a single glass of wine. Botrytized wines also come from Germany, Italy, Hungary, and other wine-growing regions around the world, but because the fungus is so unpredictable and dangerous, few winemakers are willing to take the risk and allow it to colonize their vines.

them to grow in their settlements. That selection process began about eight thousand years ago, and from there, it was simply a matter of choosing the tastiest fruit and taking cuttings to get a genetic clone. Fortunately, pottery was also being invented around the same time, leading to the happy circumstance of crushed fruit stored in a container long enough for wild yeast to find it.

One more lucky break made wine making possible. A particular species of wild yeast that feeds on the exudates of oak tree bark managed to crawl into early wine vats around five thousand years ago and do a particularly good job of fermentation. There would have been other yeasts living naturally on grape skins, but they would not have been nearly as well suited to the job. But somehow, oak yeast got into the mix.

How did this happen? Scientists have a few theories. It might be that grapevines occasionally climbed up an oak tree and picked up the yeast. It's also possible that people gathered acorns and grapes at the same time, commingling the microorganisms on each, or that insects picked up the yeast on an oak tree and carried it to a grapevine because it was attracted to the rising sugar in the fruit. However it happened, that yeast species, *Saccharomyces cerevisiae*, found its way into wine somehow. Today it is an entirely domesticated creature, rarely found in the wild and widely bred into specialized strains that are used around the world to make bread rise and to ferment wine and beer.

WARNING: DO NOT ADD WATER

During Prohibition, enterprising California grape growers kept themselves in business by selling "fruit bricks"—blocks of dried, compressed grapes that were packaged with wine-making yeast. A label warned purchasers not to dissolve the fruit brick in warm water and add the yeast packet, as this would result in fermentation and the creation of alcohol, which was illegal.

the first wine

Archeologist Patrick McGovern has analyzed ancient pottery fragments around the world and found evidence of wine making dating
back six thousand years in the Middle East. A UCLA team uncovered a complete wine production facility in Armenia from the same
period. McGovern also detected possible grape residue in pottery
fragments from 7000 BC in China. The only people who did not
develop a strong wine-making tradition with their local grapes were
Native Americans—or if they did, they've hidden the evidence very
well. South American Indians in particular made alcohol from corn,
agave, honey, cactus fruit, seedpods, and bark, but they rarely, if
ever, threw grapes into the mix.

Over time, the Egyptians, Greeks, and Romans became the world's
most sophisticated vintners. Many early scientific advances slipped
into oblivion during the Middle Ages, but wine-making technology
survived, thanks to the efforts of monks and the deep associations
between wine and religion. By the 1500s, vineyards were beginning
to transition from church enterprises to private operations, often
run by nobility. During the next few centuries, the British managed
to forget they were at war with France often enough to purchase
enormous quantities of their enemies' fine wines. Clearly, a robust
European wine market was already in place as colonists were arriving in the New World.

the invention of brandy

By that time, a tradition of distilling wine into brandy had emerged
as well. Spanish and Italian writings from the thirteenth century
show that wine was being boiled into some kind of strong spirit.
The Dutch gave it the name *brandewijn*, or "burnt wine," which was
shortened to "brandy." Dutch traders set up stills in ports where
wine was made, particularly if the wine was mediocre and would be
more profitable in the form of brandy. One such place was France's
Cognac region. The white wines made in the area were not terrible;
they were just bland. The Dutch hoped to reduce shipping costs by
distilling them into a high-proof spirit that would later be mixed
with water as a substitute for wine. Sometimes, in the confusion and
chaos of a busy port, those spirits would sit in a barrel longer than
intended. The result? Rich, complex, aged Cognac. It later became

VERMOUTH COCKTAIL

This classic cocktail is a template for experimentation with aromatized wines. Mixing Punt e Mes and Bonal Gentiane Quina, for instance, makes for a remarkably good drink. And Lillet blends well with almost anything.

1 ounce **dry white vermouth**
1 ounce **sweet red vermouth**
Dash of **Angostura bitters**
Dash of **orange bitters**
Lemon peel
Soda water (optional)

Shake the white and red vermouth and bitters over ice and strain into a cocktail glass, or serve over ice topped with soda water. Garnish with the lemon peel.

clear that even vineyard waste could be fermented: crushed skins, stems, and seeds all went back into the fermentation tank to make a high-proof spirit like grappa.

While grape brandies and eaux-de-vie were coming into their own around Europe, Spanish and Portuguese winemakers noticed that the British had a taste for sweet wines fortified with brandy. Adding extra alcohol to wine was an easy way to stop the fermentation process—yeast can't live in a higher-proof solution—but it also helps a different kind of yeast to survive. In the Jerez region in southern Spain, white wine was traditionally aged in casks that were only partially full. A particular strain of *S. cerevisiae* yeast colonized the casks and formed a thick skin over the wine. The Spanish call it *flor*; scientists call it velum. Unlike other strains of yeast, the *flor* actually prefers a higher alcohol content of around 15 percent, so winemakers would fortify the wine to keep the yeast alive.

A FIELD GUIDE *to* FORTIFIED WINES

Fortified wines are wines with higher-proof alcohol added. The most famous are:

MADEIRA: Oxidized Portuguese wine fortified with neutral grape spirit.

MARSALA: Fortified Italian wine made in the Marsala region.

MUSCATEL OR MOSCATEL: Sweet, fortified muscat wine produced mostly in Portugal.

PORT: Portuguese wine fortified with grape spirit before fermentation is over, leaving residual sugars in the blend. (In the United States, such wine made anywhere in the world can be called port, but only the Portuguese version can carry the label "porto.")

SHERRY: Spanish white wine mixed with brandy after fermentation is complete.

VINS DOUX NATURELS: Sweet, fortified French wine, often made from muscat grapes.

These wines—which the British called sherry, possibly a corruption of Jerez—are said to be biologically aged because the yeast changes the flavor over time. Adding to sherry's complexity is the use of the solera system for aging. The barrels are stacked four high, with finished sherry coming only from the bottom barrel. It is refilled by sherry from a barrel above it, which is then refilled from the barrel above that, and so on. New wine is only added to the top barrel. Some soleras have been in continual operation for over two hundred years, giving the finished product extraordinary depth and flavor.

Other regions developed their own fortified wines. Portuguese winemakers added brandy to half-fermented wine to stop the yeast from eating all the sugar. After a few years in tanks or barrels, the raisiny-sweet result is port. Madeira, also from Portugal, is made in a similar manner, usually with white wine grapes, and then exposed to air and subjected to the kinds of temperature extremes that early barrels would have encountered on long ocean voyages. This deliberate abuse gives it its oxidized, dried fruit flavor and means that it ages well and remains drinkable for up to a year after it's been opened. Italy's Marsala is similarly fortified and aged—and so it goes, in wine-making regions around the world.

Another centuries-old European tradition—that of flavoring wine with herbs and fruit—led to the invention of vermouths and aperitif wines, also called aromatized or fortified wines. They might have originally been intended as medicine—a wine infused with wormwood, quinine, gentian, or coca leaves would have represented an attempt to treat intestinal worms, malaria, indigestion, or listlessness, respectively—but by the late nineteenth century they had become respectable drinks in their own right. Vermouth is made with white wine (red vermouth is not made with red wine, but white wine sweetened and colored with caramel) and fortified slightly with brandy or eau-de-vie, bringing the alcohol content to about 16 percent.

the american experiment

With such a remarkable and diverse wine-making tradition, it must have been difficult for Europeans to set sail for a continent that might or might not have been suitable for grape growing. Early vineyards failed, which is why the Founding Fathers imported their wine or drank homemade brews made from grains, corn, apples, and molasses. Thomas Jefferson in particular spent lavishly on French wine and tried to find a native American grapevine suitable for wine making for his garden at Monticello. Neither the native nor the European varieties he planted ever produced a drop of decent wine.

What went wrong? The native varieties simply weren't suited for wine making—but more about that in a minute. The failure of the European vines was the real mystery. What Jefferson didn't know— and what no one knew until later in the nineteenth century—was that sturdy American grapevines were resistant to attacks by a tiny, aphidlike pest called phylloxera (*Daktulosphaira vitifoliae*) that was also native to America. European grapes had no such resistance, which explained why imported vines planted in American soil withered.

Before anyone realized this, however, Americans had sent a gift of native grapevines to France. Unfortunately, those vines were infested with phylloxera. They went right to work attacking the vineyards. This tiny American pest went on to devastate the French wine industry in the nineteenth century.

STRANGE RELATIONS

Carole P. Meredith, professor emerita at the University of California, Davis, Department of Viticulture and Enology, has analyzed the genetics of some of the most popular wine grapes and determined their parentage. The results? Cabernet Sauvignon is the child of Cabernet Franc and Sauvignon Blanc. An old variety called Traminer gave birth to Pinot Noir, which in turn mated with an old peasant grape called Gouais Blanc to produce Chardonnay. These couplings probably happened by accident in seventeenth-century French vineyards, well before vintners used modern plant breeding techniques.

AROMATIZED WINES REVEALED

EVEN THE MOST ADVENTUROUS WINE DRINKERS might not have explored the remarkable world of aromatized wines. These wines have herbs, fruit, or other flavors added, and might also be fortified with additional alcohol. Vermouth is the best-known example; if you don't believe that a glass of vermouth can be lovely on its own, try some of these. Just remember that, like other wines, they will spoil quickly and should be refrigerated after opening. The extra alcohol content helps them last a little longer than wine, but drink them within a month or so.

MISTELLE: A mixture of unfermented or partially fermented grape juice and alcohol, sometimes used as the base for aromatized wines. Try these:

- *Bonal Gentiane Quina:* A mistelle base flavored with gentian and quinine. Very good on its own or as a substitute for red vermouth in cocktails.

- *Pineau des Charentes:* A non-aromatized, barrel-aged mistelle with Cognac. Made in southwestern France. Unforgettable.

QUINQUINA: Fortified wines with quinine and other flavors added. Two fine examples are:

- *Cocchi Americano:* An Italian quinine, herb, and citrus-infused fortified wine used in classic cocktails but perfect on its own.

- *Lillet:* A blend of Bordeaux wines, citrus peel, quinine, fruit liqueur, and other spices. Available in blanc, rouge, and rosé styles. All three are enchanting.

VERMOUTH: Made of wine fortified with alcohol, along with wormwood, herbs, and sugar. Bottled at 14.5 to 22 percent ABV. These two will make a vermouth drinker of you:

- *Dolin Blanc Vermouth de Chambéry:* Halfway between a dry and sweet vermouth, the Dolin Blanc is a fine, balanced blend of fruit, floral, and pleasantly bitter notes. Drink it over ice with a twist of lemon.

- *Punt e Mes:* A wonderfully rich, sophisticated red aromatized wine with dried fruit and sherry flavors that is also good enough to drink on its own. Consider it a more complex and grown-up substitute for sweet vermouth.

At first, no one knew what was killing the vines. In fact, it took decades to simply understand the creature, much less find a way to kill it. Its life cycle was unlike anything scientists had ever seen. First, a generation of female phylloxera are born that never mate, never go on a single date, but are capable of giving birth anyway. The next generation is the same way, and the next, so that one generation of females is born after another. When, once a year, a batch of males finally emerge, they exist only to mate and die. The poor creatures are not even given a digestive tract; the males will not enjoy a single meal in their short, sex-filled lives. Once their job is complete, the females continue without them for several more generations. Their habitat changes, too: during one stage of their life cycle they induce the leaves to form galls—protective plant growths that hide the creatures—and during another stage they vanish underground to attack the roots.

By the time the phylloxera was finally understood, France's wine industry was nearly obliterated. Salvation came from the very plant that had caused the problem in the first place: the resilient American grapevine. Grafting fine old European vines to the rough-and-tumble American rootstock allowed the winemakers to replant and bring their industry back, although they worried that the flavor would suffer. Most wine connoisseurs would agree that French wines have done quite well in spite of the setback, but they still seek out "pre-phylloxera wines," made from those pockets of European vines that managed to survive on their own roots. Chile, for instance, produces pre-phylloxera wine because Spanish missionaries brought the grapes there, but the phylloxera never arrived.

Because wine was in short supply during the outbreak, absinthe became the drink of choice in cafes. Rumors of its toxicity are greatly exaggerated: while it is flavored with wormwood (*Artemisia absinthium*), it was never the plant itself that drove drinkers crazy. It was the extremely high alcohol content. Absinthe was bottled at about 70 percent ABV, almost twice as high as brandy. Whatever the reason for its perceived social ills, the winemakers were all too happy to join the French temperance movement in advocating for a type of prohibition that would ban absinthe but protect wine, which was seen as a healthy and moral drink.

Although the French wine industry recovered, American farmers were still trying to figure out how to make good wine from native American grapes. The difficulty had to do with the genetics of the grape itself. While the European *V. vinifera* enjoyed almost ten thousand years of selection by humans, who chose larger, tastier fruit and favored hermaphrodite vines over dioecious vines, very little human selection seems to have taken place in North America. Instead, the birds did it. They selectively picked blue-skinned varieties, an unattractive color for wine, because they could see them better—and they chose small fruit over large because they could eat it in one bite.

So even though *V. riparia*, one of the most widespread native American species, is remarkably cold-hardy and resists pests and disease, that small, blue fruit did not impress winemakers as much as it impressed the birds. After three hundred years of experimentation, American botanists are only just now figuring out how to turn native grapes into wine. University of Minnesota researchers have crossed *V. riparia* with European vines to produce new varieties, like Frontenac and Marquette, that yield surprisingly good wines, even in that cold northern climate. They are sturdy, robust and quite drinkable wines with just a hint of the wild herbaceousness that makes them uniquely American.

PISCO SOUR

This is Peru's national cocktail.

1½ ounces **pisco**
¾ ounce **fresh-squeezed lemon or lime juice**
¾ ounce **simple syrup**
1 **egg white**
Angostura bitters

Shake all the ingredients except the bitters in a cocktail shaker without ice for at least 10 seconds. The "dry shake" makes the drink foamy. Then add ice and shake for at least 45 seconds more.
Pour into a cocktail glass and sprinkle a few drops of bitters on top.

GRAPE-BASED SPIRITS AROUND THE WORLD

BRANDY is the generic term for a wine (or other fruit) spirit, usually distilled to 80 percent alcohol or less, then bottled at 35 to 40 percent alcohol. Types of grape brandy include:

AGUARDIENTE: Portuguese brandy. This term also describes neutral grape spirits.

ARMAGNAC: Made in the nearby Armagnac region. Unlike Cognac, which is made in pot stills, Armagnac is made in a continuous still called an alembic, at a lower proof. Both are made from specific varieties of grapes, then aged in oak.

ARZENTE: Italian brandy.

BRANDY DE JEREZ: This and other spirits simply labeled "brandy" come from Spain.

COGNAC: Made in France's Cognac region.

METAXA: Greek brandy.

EAU-DE-VIE is a higher-proof clear spirit made from fruit; when it is made from the pomace (skins, stems, seeds and other remnants of wine fermentation), it is called *pomace brandy*, or:

> **BAGACEIRA** in Portugal; **GRAPPA** in Italy; **MARC** in France; **ORUJO** in Spain; **TRESTER** in Germany; **TSIKOUDIA** in Greece

GRAPE-BASED GIN is any grape vodka infused with juniper and other botanicals. G'Vine is a French gin made from the same grapes used in Cognac, plus an extract of just-opened grapevine blossoms and other herbs and spices.

GRAPE VODKA is a high-proof, unaged spirit like an eau-de-vie, intended to be of a neutral character. A fine example is St. George Spirit's Hangar One Vodka, made from a blend of Viognier grapes and wheat; the grapes give it the lightest possible essence of fruit. Ciroc, made from French grapes, is another popular brand.

PISCO is named after the port city of Pisco, Peru, where eighteenth-century voyagers stopped to stock up on the local spirit. It matures in glass or stainless steel, not oak. In Peru, it is bottled at full strength, ranging from 38 to 48 percent alcohol. Chileans make a version using different grape varieties and some wood maturation.

> **ACHOLADO** comes from a blend of grape varieties.

> **MUSTO VERDE** is distilled from partially fermented grape stems, seeds, and skins.

> **PISCO PURO** is made from a single variety of grape.

POTATO

Solanum tuberosum

SOLANACEAE (NIGHTSHADE FAMILY)

On June 3, 1946, a headline in the *New York Times* read, "Potato May Avert Drinkers' Drought." Wartime grain shortages had been hard on beer and whiskey drinkers. The Agriculture Department had diverted grain to more important uses: food, livestock feed, and the production of industrial alcohol for rubber manufacturing. Restrictions continued after the war as troops wound down their operations and relief shipments to devastated postwar Europe got under way.

Because of the shortages, distilleries were allotted a single ten-day mashing period per month, with limits on the amount of rye or other grains that could go into the mash. With so little raw material to work with, the distillers got creative. They asked for a share of the nation's heavily rationed potato supply, explaining that they could put the lower-grade, smaller, misshapen potatoes to use in making blended whiskies, gins, or cordials while saving the higher-quality potatoes for food. This move, the Agricultural Department pointed out, could "change the drinking habits of Americans and make popular such potato drinks as vodka."

At the time, vodka was virtually unknown to American drinkers. In 1946, Americans drank only one million gallons of vodka, less than 1 percent of all spirits consumed in the country. By 1965, that number had grown to thirty million. Vodka has always been made of rye, wheat, and other grains in addition to potatoes, but Americans nonetheless thought of the spirit as an exotic, specifically potato-based drink.

incan treasure

The potato traces its ancestry to Peru. Wild potatoes (*Solanum maglia* and *S. berthaultii*) grew along the western coast of South America at least thirteen thousand years ago, when glaciers still covered higher elevation areas. By 8000 BC, the glaciers were receding and the coast became more dry and desertlike, so people moved up to higher elevations. It was there, in the Andes mountain range, that early Peruvians cultivated potatoes. Growing conditions were difficult and unpredictable—weather changed quickly on the rocky slopes—so thousands of different cultivars were grown, each with its own ecological niche.

The first Spaniards to encounter the Inca empire in 1528 found an astonishingly sophisticated civilization. A road system spanning more than fourteen thousand miles, highly advanced architecture, a system of taxation and public works projects, and thoroughly modern farming techniques made the Inca Empire comparable to the Roman Empire. Francisco Pizarro and his men were so dazzled by the Inca's gold and jewels that the grubby potato hardly seemed worth picking up. It was a few more decades before the potato was grown in Europe, and it was not widely cultivated as a food crop until later in the seventeenth century.

Europeans didn't trust the potato because it was a member of the dangerous nightshade family. Old World members of this family, including henbane and deadly nightshade, were highly toxic. That gave them reason to fear all the nightshades they found in the New World, including potatoes, tomatoes, and tobacco. (They were equally suspicious of eggplant, a nightshade native to India.) And in fact, the potato plant does bloom and produce small, poisonous fruit similar to that of other nightshades. Even the starchy tubers can accumulate toxic levels of the alkaloid solanine if exposed to light; this is a defensive reaction designed to protect a vulnerable, unearthed potato from predators.

Because it was a nightshade, and because it was eaten by so-called primitive people in South America, potatoes were seen at best as a commodity to be fed to slaves and at worse as a dirty, evil root that caused scrofula and rickets. The fact that the Irish embraced the potato only helped convince the English that it was a lowly food fit only for a peasant. Nonetheless, it did eventually become established throughout Europe. Explorers also took it to Asia, Africa, and to the new colonies in North America.

the birth of vodka

Ask people today about the invention of vodka, and you might hear that it is made from potatoes and that it comes from Russia. Neither statement is entirely true. Vodka was already being distilled from grains long before the potato ever arrived in Europe. The birthplace of vodka is the subject of endless dispute between Russia and Poland, with each claiming the spirit as its own. What is known for certain is that a clear, high-proof spirit distilled from grain was made throughout the region by the 1400s. The Polish term *wodki,* meaning "little waters," was used by Stefan Falimirz in his 1523 medical text *On Herbs and Their Powers,* long before potatoes could have been used in *wodki.* They were only just being discovered in Latin America at that time and had not yet reached Europe.

By the eighteenth century, potatoes were a staple food crop in eastern Europe, and distillers were experimenting with them as early as 1760. Those early trials must have been difficult. Potatoes, after all, are merely thickened stems that grow underground and store food and water for the next generation. Unlike the starch in grain, the starch in a potato is not intended to be converted to sugar all at once to feed a germinating seed. Instead, it is released slowly, over a long growing season, to nourish a young plant. This is a brilliant survival strategy for the potato, but it doesn't help the distiller.

A Polish pamphlet called *The Perfect Distiller and Brewer,* published in 1809, described the process of distilling vodka from potatoes, with the warning that it was the worst kind of vodka, behind vodka made from sugar beets, grains, apples, grapes, and acorns. In fact, potatoes only became a common ingredient in Polish vodka because they were cheap and abundant, not because they made a high-quality spirit. They tended to turn into a thick, sticky paste in the fermentation tank, the starch was not easy to convert to sugar, and they produced higher levels of toxic methanol and fusel oils. Russian vodka makers looked down on cheap Polish potato vodkas; to this day, they insist that the best vodka is made from rye or wheat instead.

⊰ SWEET POTATO ⊱

Ipomoea batatas

CONVOLVULACEAE (MORNING GLORY FAMILY)

THE SWEET POTATO IS NOT REALLY A POTATO AT ALL—it is the root of a climbing vine closely related to the morning glory. And, by the way, it has nothing to do with the yam, which is a starchy root in the *Dioscorea* genus grown in Africa. (While Americans have traditionally called soft, orange sweet potatoes yams, true yams are almost never sold in the United States.)

Sweet potatoes are native to Central America and traveled around the world courtesy of European explorers. One of the earliest alcoholic beverages made from the spirit was mobbie, a fermented drink of sweet potatoes, water, lemon juice, and sugar, which was described in Barbados as early as 1652. It was a popular "small beer" for over a century, until a plague of sweet potato beetles wiped out the crop. Sugarcane plantations took over sweet potato fields and rum became the drink of choice.

Brazilians also made a fermented drink of the tubers, called *caowy*. It wasn't much to Europeans' liking: writing in 1902, American winemaker Edward Randolph Emerson said that the Portuguese improved the drink's flavor by renaming it *vinho d'batata*, "which sounds much better and sometimes there is a lot in a name."

The best-known sweet potato spirit is Japanese *shochu*, a distilled beverage of up to 35 percent alcohol that can be made from sweet potatoes, rice, buckwheat, and other ingredients. Korean *soju* is also sometimes made of sweet potatoes.

Throughout Asia, "sweet potato wine" refers to a homebrew not too different from what the islanders drank in Barbados. A sweet potato beer is made in North Carolina and in Japan, and sweet potato vodkas are just coming on the market.

artisanal potatoes

By the time American distillers were applying to use surplus potatoes for their whiskey blends in 1946, vodka was poised for a comeback. The troops returning home from Europe had done a little drinking in foreign lands. They were ready to try something new. With the postwar prosperity came a new era in cocktail drinking. Mixed drinks like the Moscow Mule and the Bloody Mary won over drinkers who liked vodka as a neutral, all-purpose mixer. Whether it was made of grain or potatoes didn't matter so much. During the last half of the twentieth century, vodka became the spirit of choice for cocktails.

Now potato vodka is getting another boost from foodies' enthusiasm for artisanal vegetables. Chopin, a Polish potato vodka launched in North America in 1997, quickly became a popular premium brand. Many other Polish vodkas followed suit. Craft distillers in Idaho, New York, British Columbia, and England are selecting specific potato varieties the way a winemaker chooses a grape and rolling out their own locally made vodkas.

But does the variety of potato really make a difference? There's not much consensus among distillers on this point. Potato vodkas have what is described as an oily, full-bodied taste as compared to grain vodkas, but whether you can taste the Russet Burbank or the Yukon Gold is between you and your palate.

Tyler Schramm of Pemberton Distillery in British Columbia uses a blend of five potato varieties, but he selects them more for their starch content than flavor. "I did my master's thesis on potato distillation," he said, "and I tried single-variety distillations. Our vodka is a sipping vodka, so it's meant to have some flavor. But there is no flavor difference from one potato variety to the next that anyone could really pick out." Of greater importance to him is environmental stewardship and the distiller's traditional role in putting otherwise unusable food to use. A single bottle of his Schramm vodka requires fifteen pounds of potatoes, so he only wants to take the part of the crop that could not go to feed people. To do that, he purchases from organic farmers and asks for misshapen or oddly sized potatoes that the farmer would not otherwise be able to sell.

He believes that the climate in British Columbia also works to his advantage: "Unlike grains, potatoes do not store well," he said. "We can do this in our cold climate, but it won't work everywhere."

Karlsson's Gold vodka, made in Sweden, is distilled from a carefully chosen blend of seven varieties: Celine, Gammel Svensk Röd, Hamlet, Marine, Princess, Sankta Thora, and Solist. The vodka is distilled only once and bottled with minimal filtration, so that the flavor of the potatoes comes through. Master blender Börje Karlsson, who also created Absolut, believes he has produced a vodka that should be savored on its own. "Just drink it as it is," he said forcefully in an interview. "If you don't like it, don't drink it." In fact, their signature drink, the Black Gold, is surely inspired by a baked potato. All it needs is a little butter.

BLACK GOLD

1½ ounces **Karlsson's Gold vodka**
Cracked black pepper

Fill an Old-Fashioned glass with ice cubes and pour the vodka over the ice. Crack black pepper over the ice.

 RICE

Oryza sativa var. *japonica*

POACEAE (GRASS FAMILY)

For such an ancient and important plant, rice has not figured prominently in the tastes of American drinkers. In 1896, the *New York Times* called sake a "vile rice wine" and said that it had a "markedly poisonous effect" on native Hawaiians, who were choosing it over "less unwholesome California wines."

Even today, we tend to think of sake as a miserable hot, sour, yeasty drink we once tasted at the urging of an aunt who took us to a Japanese restaurant in Kansas City. But making a decision about sake based on a bad memory of that cheap, low-grade *futsu-shu* would be like judging wine based on a jug of Boone's Farm. In fact, sake is as diverse and interesting as wine, with an even longer history. And just as the grape is made into an endless parade of spirits, rice has been put to use in a wide range of alcoholic beverages around the world. It turns up in Budweiser, it's an ingredient in premium vodkas, and its surprisingly floral essence is captured in Japanese *shochu*.

no ordinary grass

Evidence uncovered by both archeologists and molecular geneticists points to China's Yangtze Valley as the origin of all varieties of rice grown around the world. It was domesticated there between eight thousand and nine thousand years ago. Making some sort of drink out of it was clearly the first order of business: archeologist Patrick McGovern found evidence of an eight-thousand-year-old brew of rice, fruit, and honey at the Jiahu site in Henan Province. (He worked with Dogfish Head brewery to re-create the brew, which they named Chateau Jiahu.) It would take centuries of trial-and-error to develop the intricate process used to create modern sake, but those early rice wines were headed in that direction.

But first, rice diversified and spread around the world. It is a water-loving grass that reaches up to sixteen feet in flooded fields. However, it does not have to grow in standing water. Its peculiar method of cultivation in rice paddies probably came about when people noticed healthy rice plants growing in flooded fields during the monsoon season. The plants just happen to have a well-developed system of airways that carry oxygen from the tip of the leaves down to the roots, just like aquatic plants do. Without this, they would rot and die during a flood. But unlike aquatic plants, they can grow in regular soil as well.

Growing rice in flooded fields turned out to be a useful strategy for early farmers throughout Asia and India. Low-lying areas prone to flooding were useless for any other crop but perfect for rice. The flooded fields were blissfully weed-free, since terrestrial weeds are incapable of living in standing water whereas the aquatic varieties cannot survive when the floods recede.

Rice is wind-pollinated and wildly diverse; over millennia, new varieties have been selected not just for their flavor and size but also for their ability to tolerate specific soil types and water levels and for the extent to which the grains cling to the stalk after they are ripe so they can be harvested. There are over 110,000 different varieties of rice around the world—and this does not include so-called wild rice, *Zizania* spp., a related grass native to North America and Asia. For the purposes of making alcohol, just a few specialized varieties of *Oryza sativa* var. *japonica* take center stage. But the rice is only part of the story. To understand how rice becomes a drink like sake, you have to understand the mold.

sake

As with any grain, fermentation cannot begin until the starches have been converted to sugar. This can happen all by itself by getting the grain wet, which encourages enzymes to turn starch to sugar to feed the emerging seedling. Brewers could speed this process up with malted barley, which possesses abundant levels of those enzymes. But Asian cultures found other ways to do it. The Japanese method is just one example, but it's the best known. First, the rice is first milled to remove some of the outer coating, called the bran. The bare, brown grains have to be carefully polished to strip the bran away without crushing the rice. Leaving each grain intact is tricky: corn, oats, wheat, and other grains are often milled and crushed at the same time to make meal or flour, so it takes a different approach to mill rice without breaking it at the same time.

The technology used to polish rice has changed little over the centuries. Although the equipment is more sophisticated, it still involves passing grains of rice across an abrasive stone hundreds of times to grind away the outer coating, leaving only a pristine, white kernel of starch. The only difference is that today's machines have more endurance than human-powered mills. A modern brewery might polish its rice for four days straight, resulting in a remarkably smooth and even removal of the bran. The quality of sake is thought to be vastly better today than it was a hundred years ago; sophisticated milling technology gets most of the credit.

The variety is also important, just as the grape variety is to wine making. In good sake rice, the nutrients are not distributed throughout the grain. Instead, it has a kernel of pure starch inside and nutrients on the outside, which means that it can be more easily polished away. Yamada Nishiki is the best-known high-end rice variety for sake brewing; it was bred in the 1930s from two older strains of sake rice and is considered a full, round, mellow-flavored rice. Other rice varieties include Omachi, prized for its wild herb and floral flavors, the cold-tolerant Miyama Nishiki, and Gohyakumangoku, which was developed in the 1950s for lighter, machine-made sakes. On the West Coast, the ubiquitous Calrose rice, developed in California in 1948, is used by Sake One, a brewery just outside Portland, and other American sake makers.

Even more important than the variety of rice is the extent to which it is polished. This is the way to judge a good sake; the finest styles are made of rice that has been polished to half its original size. This gives the mold—which we'll get to in a second— less protein, oils, and nutrients to contend with. It can go straight to the starchy core of the rice and do its work.

ENJOYING SAKE

Good sake should never be served hot. The tradition of heating sake was a way to hide the taste of rough, poorly made sake. Better fermentation technology has led to higher-quality sake that almost always tastes better cold. Drink it fresh: most sake brewers advise against storing a bottle for more than a year. Once opened, it lasts in the refrigerator slightly longer than wine, but it should be finished off within a couple of weeks. Because there are so many styles of sake, the best way to get acquainted is to go to a sake bar with some friends and order a tasting.

The polished rice is washed, soaked in water, and sometimes steamed, all of which helps increase the moisture content. At that point it's taken into a room that resembles a Japanese sauna—warm, cedar-lined, and extremely dry. The damp rice is spread out in an enormous bed, and there it meets the mold, a species of fungus called koji, *Aspergillus oryzae.* Koji was domesticated in China about three thousand years ago; it traveled to Japan a thousand years later. Like *Saccharomyces cerevisiae,* the yeast used in the West for fermentation and bread making, koji is an entirely domesticated creature now. In addition to its role in sake production, koji is used to ferment tofu, soy sauce, and vinegar, making it a sort of staple microorganism in Japanese cuisine.

The koji mold spores are sprayed on top of the bed of damp rice. Normally mold would simply grow on the surface—picture a moldy loaf of bread—but the dry atmosphere forces the mold to grow into the bed of rice, and into each kernel of rice, to seek the moisture it needs to survive. There, inside that damp, starchy bit of grain, it releases the enzymes that break starch apart and turn it into sugar.

At the same time, a separate batch of rice, koji, water, and yeast are blended together to kick-start fermentation. The koji only turns the starch to sugar; now yeast have to eat the sugar and turn it into alcohol. Once the yeast start multiplying, the two batches are combined over three or four days, with more steamed rice, water, and koji meeting the yeast each day. At that point the two processes are

NO. 1 SAKE COCKTAIL

In the last few years, Asian restaurants in the United States have felt some obligation to create cocktails from sake and *shochu*. This is a shame, because both drinks are lovely on their own and seem to resist mixing—the flavors just don't marry well with other cocktail ingredients. But here, after much experimentation, is one sake cocktail that is a proven crowd-pleaser. It's easy to make a batch before a party, which is why it is presented in portions rather than ounces. Make as much or as little as you need.

4 parts *nigori* (unfiltered) sake
2 parts mango-peach juice (a bottled blend is fine)
1 part vodka
Dash of Domaine de Canton ginger liqueur
Drop of celery bitters

Mix all the ingredients except the bitters briskly; then taste. It might need more ginger liqueur or vodka at this stage. Keep chilled until your guests arrive and then pour into cocktail glasses. Add a drop of celery bitters to the top of each drink as you serve it.

happening simultaneously in the same vat: the koji mold is breaking the starch into sugar, and the yeast is eating the sugar as it is released. This complex brew of microbes has to be managed very carefully and stopped at the precise moment the sake is finished. Because each ingredient is added gradually, the yeast don't die off as quickly as they do in, say, wine or beer fermentation. They continue to live in the mash and excrete alcohol until the alcohol content reaches about 20 percent.

Once the brewer is satisfied, the entire yeasty, moldy mash is pressed to separate the sake from the solids. Then it is filtered and pasteurized with heat to stop the fermentation. Some enzymes survive and continue to work on the brew, so the flavor improves as it matures in tanks for a few more months. While most sakes are sold clear and full strength, some are diluted to make the alcohol content closer to that of wine, and some are only coarsely filtered so that the resulting beverage is cloudy from the remnants of yeast, koji, and bits of undigested rice. A high-quality sake will taste clean, crisp, and bright, with aromas of pears and tropical fruit or, in some cases, an earthier, almost nutty aroma.

rice spirits

Those distinctive sake flavors are even more concentrated in *shochu,* a distilled drink that starts with a sakelike mash. It's bottled at only about 25 percent alcohol, and loopholes in some United States liquor laws allow it to be served in restaurants with only a beer and wine license. This has led to *shochu*'s use as a mixer in Asian-inspired cocktails—think lemongrass martinis—but it's actually best on its own, over ice. *Shochu* is also made from barley, sweet potatoes, buckwheat, and other ingredients, but the rice-based version is the most common. And "common" is an understatement: the best-known *soju* brand (the Korean version of *shochu*), Jinro, outsells every other spirit brand in the world, with the possible exception of some Chinese brands that don't disclose their sales. More Jinro *soju* is sold each year than Smirnoff vodka, Bacardi rum, and Johnnie Walker whiskey combined—608 million liters in all.

Drinks similar to shochu and sake can be found throughout Asia. In addition to Korean soju, a Chinese rice wine similar to sake is called *mijiu.* In the Philippines, rice wine is *tapuy* and in India it is *sonti.* In Bali they make *brem,* in Korea a sweet version is called *gamju,* and in Tibet it is *raksi.*

Fermented rice cakes are also added to water to make home brews throughout Asia. One of the most interesting uses of rice cakes was described by French anthropologist Igor de Garine, who did field work in the Malaysian state of Terengganu in the 1970s. As deeply devout Muslims, the villagers he lived with never touched alcohol. But they did have a tradition of making steamed rice cakes called *tapai.* The cakes were cooked in combination with local yeast, wrapped in the leaf of a rubber tree, and left out in the heat for a few days. They fermented so well that when he tasted one, he thought that "someone had slipped a little gin into it." He never mentioned the familiar flavor to his hosts, who managed to get some enjoyment from the cakes without realizing—or acknowledging—that they contained alcohol.

Rice is not limited to sake, *shochu,* and fermented rice cake. Kirin and many other Japanese beers are made with rice, as is Budweiser and a few other American beers. Premium rice-distilled vodkas have come onto the market in the last few years. At the other end of the spectrum, a Laotian rice whiskey called *lao-lao* is touted as the cheapest spirit in the world, at only about a dollar a bottle—and that bottle includes a perfectly preserved snake, scorpion, or lizard, a gimmick that puts the worm in mezcal to shame.

SAKE NOMENCLATURE

DAIGINJO: The highest-quality sake, with at least 50 percent of the grain polished away.

GINJO: The next highest designation; at least 40 percent of the grain is removed.

JUNMAI: No particular level of milling required, but the percentage must be stated on the bottle.

- -

GENSHU: Full-strength sake, up to 20 percent alcohol.

KOSHU: Aged sake (uncommon).

NAMA: Unpasteurized sake.

NIGORI: Cloudy, unfiltered sake. Shake before serving.

RYE

Secale cereale
POACEAE (GRASS FAMILY)

R ye was an unlikely candidate for domestication. The grains are rock hard and not particularly tasty to livestock. Yields are low. It is guilty of "precocious germination," which means that the seeds can sprout while they're still on the stalk. At worst, that spoils the grain, and at best, it makes it impossible for brewers and bakers to work with, because once the conversion of starch to sugar gets under way, the carefully controlled process of getting bread to rise or turning grains to alcohol is thrown off course.

Rye also happens to be low in gluten and high in a carbohydrate called pentosan. As compared to wheat, the proteins in rye are very water soluble, which means that they turn into a slimy liquid or a rubbery solid when wet. This makes dough less elastic and can turn a brewer's mash into a sticky, awful mess. Most rye dough has to be mixed with wheat flour to make it easier to work with; distillers limit the amount of rye in their brews for the same reason.

Pliny the Elder was not fond of rye. In his *Natural History*, written around 77 AD, he wrote that it was "a very inferior grain, and is only employed to avert positive famine." He said that it was black and bitter and had to be mixed with spelt to make it palatable but was still "disagreeable to the stomach."

This may explain why rye was one of the last cereal crops to be domesticated. It's only been cultivated since 500 BC, and even then it only became popular in Russia and eastern and northern European climates, where its cold-hardiness made it a grain of last resort. The seeds can germinate when soil temperatures hover just above freezing, allowing it to be planted in late fall, and it will survive

NOT TO BE CONFUSED *with* RYEGRASS

Ryegrass is a grass in the *Lolium* genus, unrelated to cereal rye, that is planted for erosion control and grazing. It a major cause of seasonal allergies. One species, darnel (*L. temulentum*) looks very similar to wheat and invades wheat fields. It is also host to a poisonous fungus, *Acremonium*, which causes "ryegrass staggers" in cattle.

long, harsh winters and produce a crop in spring, before any other grain. It crowds out weeds and thrives in poor soil where little else will grow.

It's no wonder, then, that European settlers brought rye with them to the American colonies. Wheat proved difficult to grow in New England's short growing season, but rye could make it through the inhospitable winters. Early American whiskies were made from whatever grains were available—usually a blend of rye, corn, and wheat.

the founding distiller

George Washington is America's most famous early distiller of rye whiskey. Like many of the nation's founders, he made his living as a farmer. In 1797, less than a year after the conclusion of his second term as president, he built a distillery at the urging of his farm manager, a Scot named James Anderson. Anderson pointed out that Washington owned the entire supply chain: he grew and harvested the grains on his own land, ground them into flour or meal at his own gristmill, and could easily transport his products to market. Converting those grains to whiskey would be the most profitable way to sell them, and Anderson had the experience to make it happen.

Washington's whiskey was a blend of available grains; a typical recipe was 60 percent rye, 35 percent corn, and 5 percent barley. It was not bottled or labeled but sold by the barrel as "common whiskey" to be dispensed to customers at taverns in nearby Alexandria. The venture was wildly successful: when he died in 1799, it was one of the largest distilleries in the country, producing over ten thousand gallons of alcohol in a single year.

After Washington's death, the distillery fell into disrepair and burned down in 1814. Fortunately, the Distilled Spirits Council of the United States took an interest in the historical site. Working with archeologists and the Mount Vernon estate, they financed the reconstruction of the distillery. It sits next to the gristmill and now operates as a working distillery, producing rye whiskey using the same equipment and methods James Anderson would have used. There's just one difference: the whiskey sold at Mount Vernon today isn't sold as unaged "common whiskey." It's aged in oak barrels to make it more palatable, then bottled and sold in limited quantities each year.

rye makes a comeback

Rye whiskey can have an edge to it. Pliny described the flavor as bitter, but it might be more accurate to call it spicy or robust. While rye whiskey might have once had a reputation as a cheap, rotgut spirit, sophisticated distillation techniques and the wondrous quality of wood maturation mean that some of the best whiskies on the market today are predominantly rye.

RYE WHISKEY: TO EARN THE LABEL "RYE WHISKEY" IN THE UNITED STATES, THE SPIRIT MUST CONSIST OF AT LEAST 51 PERCENT RYE, BE DISTILLED TO NO MORE THAN 80 PERCENT ALCOHOL, AND BE AGED IN CHARRED NEW OAK CONTAINERS AT AN ALCOHOL LEVEL OF NO MORE THAN 62.5 PERCENT. IF IT HAS BEEN AGED FOR AT LEAST TWO YEARS, IT CAN BE CALLED "STRAIGHT RYE WHISKEY."

The grain is also an ingredient in some German and Scandinavian beers, and American craft distillers are using it as well. Rye has always been a base ingredient for Russian and eastern European vodkas; now American vodka distillers use it, too. Square One vodka is made entirely from organic Dark Northern and other North Dakota rye varieties. Rather than seek out the more flavorful bread-quality varieties, the distillery tends to choose ryes based on starch content,

MANHATTAN

The Manhattan is a classic cocktail that puts rye whiskey to its highest and best use, with the sweet vermouth playing off the bitter bite of the rye. It is also a template for endless variations: replace the rye with Scotch and you've got a Rob Roy; replace the vermouth with Benedictine and you've got a Monte Carlo; or just swap sweet vermouth for dry, and garnish with a lemon twist to make a Dry Manhattan.

1½ ounces rye
¾ ounce sweet vermouth
2 dashes Angostura bitters
Maraschino cherry

Shake all the ingredients except the cherry over ice and strain into a cocktail glass. Garnish with the cherry.

and some of those are the same varieties grown as cattle feed. "All we really want is the starch molecule," said owner Allison Evanow. "The nutty flavor is not as important in a distilled clear spirit." One problem she encountered: bugs. "If they grow it for feed, it's normal to leave more bugs in it," she said. "We had to reject one supplier because there were too many grasshoppers in the grain."

Rye growers face another challenge: the grain is vulnerable to a fungus called ergot (*Claviceps purpurea*). The spores attack open flowers, pretending to be a grain of pollen, which gives them access to the ovary. Once inside, the fungus takes the place of the embryonic grain along the stalk, sometimes looking so much like grain that it is difficult to spot an infected plant. Until the late nineteenth century, botanists thought the odd dark growths were part of the normal appearance of rye. Although the fungus does not kill the plant, it is toxic to people: it contains a precursor to LSD that survives the process of being brewed into beer or baked into bread.

While a psychoactive beer might sound appealing, the reality was quite horrible. Ergot poisoning causes miscarriage, seizures, and psychosis, and it can be deadly. In the Middle Ages, outbreaks called St. Anthony's fire or dancing mania made entire villages go crazy at once. Because rye was a peasant grain, outbreaks of the illness were more common among the lower class, fueling revolutions and peasant uprisings. Some historians have speculated that the Salem witch trials came about because girls poisoned by ergot had seizures that led townspeople to conclude that they'd been bewitched. Fortunately, it's easy to treat rye for ergot infestation: a rinse in a salt solution kills the fungus.

SORGHUM

Sorghum bicolor

POACEAE (GRASS FAMILY)

O n February 21, 1972, President Nixon, his staff, and members of the American media attended a banquet in Peking to mark the beginning of Nixon's historic trip to China. The ceremonial drink that night was *mao-tai,* a sorghum spirit with an alcohol content over 50 percent. Alexander Haig had sampled the drink on an advance visit and cabled a warning that "Under no repeat no circumstances should the President actually drink from his glass in response to banquet toasts." Nixon ignored the advice and matched his host drink for drink, shuddering but saying nothing each time he took a sip. Dan Rather said it tasted like "liquid razor blades."

Mao-tai is just one category in the broad class of Chinese sorghum liquors known as *baijiu.* Other grains can be used—millet, rice, wheat, barley—but sorghum has a long history in Asia, where the earliest distilled spirits from the grain were made two thousand years ago.

survival of the fittest

Why sorghum? It isn't the flavor, certainly: *baijiu* and sorghum beers aren't winning many medals from tasting panels. But sorghum happens to be incredibly drought-tolerant and easy to grow in poor soils. It can wait out periods of stress and bounce back quickly. A thin waxy cuticle keeps the plant from drying out, and natural tannins protect it from insect attacks. Young shoots produce cyanide in response to drought, which is deadly for livestock but protects the plant during a critical time.

Sorghum is, in short, a survivor. This makes it the grain of famines and poverty. It has kept people alive in times when nothing else would grow. Its simple presence in highly populated, poverty-stricken areas around the world make it the default grain for home brews.

Sorghum and millet are often mentioned in the same breath; the reason for this is that "millet" is a catchall term for at least eight different species of grain, including sorghum, that produce panicles, or small seeds in loose clusters. Some millets are called broom-corn; the broom shape is an apt descriptor. Like most millets, sorghum is a dense, tough grass that can grow to fifteen feet.

It originated in northeast Africa around Ethiopia and the Sudan and was domesticated in 6000 BC. Because it was such a useful food source, it spread across Africa, making it to India over two thousand years ago. From there it went to China along silk trading routes. There are over five hundred varieties, broadly categorized as sweet sorghum or grain sorghum. For the purpose of making alcohol, sweet sorghum is better for pressing sugar from the stalks to distill a drink like rum, and grain sorghum is better for beer or whiskey.

The grain is not particularly useful for bread making because it lacks the gluten that allows dough to stretch and rise, but it is used to make traditional flatbreads. The mouthwatering Ethiopian *injera* is made from sorghum or teff, another milletlike grain.

Sorghum's main advantage is that it is high in fiber and B vitamins, supplying much-needed nutrients when food is scarce. Once corn became widespread, this was especially important: an exclusively corn-based diet can cause pellagra, a dangerous and even deadly B vitamin deficiency. Eating corn in combination with sorghum prevents pellagra.

sorghum beer

The most practical use of the grain is as a porridge or gruel; for this reason, the first fermented drinks made from sorghum were simply thin porridges allowed to sit for a few days until the alcohol content reached 3 or 4 percent. Traditional African sorghum beers are made today in much the same way they were for thousands of years. The stalks are cut and the grains threshed by beating them on a wooden platform or a grass mat, and then the grains are soaked for a day

THE WORLD'S MOST-IMBIBED PLANT?

QUICK: WHAT PLANT TURNS UP IN MORE COCKTAILS, BEERS, AND WINES THAN ANY OTHER? Barley is a good candidate, and so are grapes. But sorghum's use in alcoholic beverages is so widespread in Asia and Africa that it may be in the running as well. Statistics are hard to come by because so much Chinese *baijiu* and African sorghum beer is made at home, often in remote rural areas, but consider this: Official Chinese production of *baijiu* is reported to be 9 billion liters per year, but homemade stills could easily produce a few billion more—and that doesn't include Chinese beer made with sorghum. (China is the world's largest beer market, consuming about 40 billion liters, nearly twice what the United States drinks.)

Then there's Africa: The amount of sorghum-based beer drunk in African nations each year is conservatively estimated at 10 billion liters, but other estimates put it as high as 40 billion. China and Africa alone, then, probably drink at least 20 billion to 40 billion liters of sorghum-based beer and spirits. And that doesn't include the sorghum used in commercial brews in the rest of the world.

Consider that global wine consumption comes to about 25 billion liters a year, with brandy and other grape-based spirits adding perhaps another billion or two. Beer drinkers put away about 150 billion liters a year, and grain-based whiskey and vodka add another 9 or 10 billion—but those are made from a mix of grains, including sorghum. So grapes, and grains like barley and rice, are clearly in the running. But if we had a way to accurately tally the world's vast and complex drinking practices, sorghum is clearly one of the world's most-imbibed plants.

or two to start the germination process. They are spread out, usually on a mat of green leaves, and covered so they will germinate for a few more days. Enzymes in the grain get to work converting starches to sugar. The malted grains are combined with hot water and ground sorghum, and then are allowed to cool. After a few days of natural fermentation, it might be brought to a boil again and allowed to cool, then more malt is added and the fermentation continues for a few more days. Once the beer is ready, it is only slightly filtered, resulting in a cloudy or opaque drink.

Making sorghum beer is often women's work. International aid groups are reluctant to discourage the practice because it brings in a little money and does give the family some nutrition. Children are given the dregs of sorghum beer to drink; the thick, yeasty remains are low in alcohol, generally free of harmful bacteria, and high in nutrients. In fact, the only real danger presented by sorghum beer comes from the containers used to brew a batch. Some Africans are genetically predisposed to iron overload, and the iron kettles or drums used for brewing, combined with the iron naturally found in sorghum, can result in beer that is dangerously high in iron for them. Unwashed containers that previously held pesticides and other chemicals have also been implicated in accidental beer-related poisonings, but the beer itself was not to blame.

Fifty years ago, this kind of homemade beer represented 85 percent of all alcohol consumed on the African continent, but that's changing quickly. Premade ingredients—sorghum flour, yeast packets, brewing enzymes—are cheap and widely available, as are "just add water" beer mixes. One step up from homemade brews is Chibuku, a fresh sorghum beer sold in cartons. The beer continues to ferment in the carton, so it has to be vented to allow carbon dioxide to escape; otherwise it would explode. One brand, Chibuku Shake-Shake, has been purchased by the global beer conglomerate SABMiller, showing just how much money there is to be made from cloudy, sour sorghum beer.

In fact, SABMiller is working to put sorghum to better use as a beer ingredient. The company has contracted with thousands of farmers in South Africa and other African nations to grow sorghum for its breweries. This allows the company to make bottled "clear beers" that resemble Western-style beers and sell them locally for less than a dollar each. Home brews can still be made for pennies per serving, but beer companies hope that Africans with just a few dollars to spend will spend some of it on higher-quality beer.

american sorghum

The grain is widely grown in the American South; in fact, it's the fourth largest crop in the nation, behind corn, wheat, and soybeans. Some form of broomcorn was grown during in the eighteenth century, but sorghum as we know it today wasn't grown here until a remarkable set of experiments began in 1856, when the editor of *American Agriculturist* magazine planted a seventy-five-foot row of sorghum from seed he imported from France. The crop he harvested—sixteen hundred pounds in all—went out in small packets to his thirty-one thousand subscribers. He repeated the stunt two years later. The United States Patent Office distributed large quantities as well, including varieties sourced from China and Africa. With free seeds arriving by mail, it didn't take long for farmers to start growing it as a fodder and grain crop—and then they realized it made good moonshine, too.

In 1862, *American Agriculturist* ran an advertisement for "sorghum wine," made from syrup pressed from the sweet stalks, which was touted as being "difficult to distinguish from the best Madeira wines." North Carolina governor and U.S. senator Zebulon Vance, reflecting on his days as a Confederate officer in the Civil War, remembered a drink made from sorghum "sugarcane." He said that "in its flavor and in its effects it was decidedly more terrible than 'an army with banners,'" meaning that it was worse than enemy fire. Which is not to say that Vance was opposed to homemade hooch. He disapproved of whiskey taxes and revenue agents chasing after moonshiners; in 1876, he complained that "the time has come when an honest man can't take an honest drink without having a gang of revenue officers after him."

Sorghum syrup liquor continued to be produced illegally. In 1899, moonshiners in South Carolina were arrested for making a sorghum spirit called tussick, which probably got its name from "tussock," a clump of grass. It was also called swamp whiskey for the swampy water that went into it. North Carolinian moonshiners

HONEY DRIP

This recipe, named after a popular sweet sorghum cultivar, is dessert in a glass.

½ ounce **sorghum syrup**
1½ ounces **bourbon**
   (or if you don't like bourbon, try it with dark rum)
½ ounce **amaretto**

Because sorghum syrup can be too thick to easily pour or measure, try spooning it into a measuring cup and heating it in the microwave for 10 seconds with a very small amount of water, just enough to make it easy to pour. (Alternatively, drop a dollop of the syrup in the cocktail shaker and hope for the best.) Shake all the ingredients over ice and serve in a cocktail glass.

referred to their sorghum cane spirit as monkey rum; while the term has disturbing racial overtones, some writers at the time claimed that it was so named because drinking it made a person want to climb a coconut tree.

Sorghum moonshining continued well into the twentieth century. In 1946, when postwar grain shortages posed a problem for moonshiners and legal distilleries alike, a four-thousand-gallon still in Atlanta exploded, and the fire destroyed three thousand gallons of sorghum syrup intended for distillation. In 1950, 789,000 tons of sorghum were used to make legitimate distilled spirits, but that number dropped to just 88,000 tons by the 1970s, the last time statistics were kept. From the 1930s to the 1970s (the only years the numbers were published), more sorghum was distilled than rye.

In spite of our long tradition of making spirits from sorghum—and in spite of the fact that even now American farmers produce four to six million gallons of sorghum syrup, there are very few sorghum spirits on the market today. In 2011, Indiana-based Colglazier & Hobson Distilling began production of a sorghum syrup rum, which they're calling Sorgrhum. (It can't legally be called rum, because rum, by law, can only be made from sugarcane, so the decidedly un-romantic name "sorghum molasses spirit" or "spirit distilled from sorghum molasses" must appear on the label instead of "rum.") The Old Sugar Distillery in Madison, Wisconsin, makes small batches of Queen Jennie Sorghum Whiskey. This, in addition to the sorghum beer being marketed to gluten-intolerant beer drinkers, may repre-sent the beginning of a sorghum revival in the United States.

an international incident

Sorghum grain has also been an important beer ingredient in China, and the Chinese learned to press the stalks of sweeter varieties to extract the juice and make wine. But the distilled hooch known as *baijiu* is China's best-known sorghum drink. The style that Presi-dent Nixon drank, *mao-tai,* is said to have originated in the Guizhou Province over eight hundred years ago. The story—which is often repeated but impossible to verify—is that *mao-tai* was sent to the 1915 Panama-Pacific International Exposition in San Francisco. A Chinese official, worried that the national product was being overlooked, dropped a bottle and shattered it, allowing the smell to permeate the exhibit hall. That got people's attention, and it won a gold medal. (Unfortunately, no records of this incident, or of the gold medal, survive in the exposition's archives.)

Mao-tai, and particularly a premium brand called Moutai, is the beverage of choice for banquets and celebrations. It made the news in early 2011 when prices hit two hundred dollars per bottle in China while selling for half that in Europe and the United States. The distillery is state run, so the high prices caused protests by citizens who felt their national drink should be more affordable to them. (Meanwhile, of course, people brew their own in homemade stills.) Although government secrecy makes Chinese markets notori-ously difficult to analyze, liquor industry experts believe that if the most popular *baijiu* brands reported their sales, they would easily outstrip the world's other top-selling brands, including the current leader, Jinro *soju,* and other popular brands like Smirnoff vodka and Bacardi rum.

The *mao-tai* served to President Nixon was surely the best China had to offer. At the state dinner, Prime Minister Chou En-lai held a match to his glass to show the president that the spirit could be lit on fire, a fact Nixon filed away for future use. In 1974, National Security Advisor Henry Kissinger told another Chinese official that the president tried to repeat the trick for his daughter when he returned home. "So he took out a bottle and poured it into a saucer and lit it," Kissinger said, "but the glass bowl broke and the *mao-tai* ran over the table and the table began to burn! So you nearly burned down the White House!"

BEWARE THE WITCHWEED

SORGHUM IS VULNERABLE TO ATTACKS by a bizarre parasitic plant called witchweed, or *Striga* spp. (Fans of the Italian liqueur Strega will recognize this Latin word for "witch.") The seeds of witchweed can only germinate in the presence of strigolactone hormones given off by the roots of sorghum. Once the seeds encounter that hormone, they send out a tiny, hairlike structure to penetrate the roots. Soon they colonize the roots, so that by the time the striga plant emerges aboveground, the sorghum is mostly dead.

Striga flourishes alongside its dying host, producing beautiful red flowers while the sorghum turns yellow and dies. An individual striga can produce fifty thousand to half a million seeds, enough to decimate a sorghum crop. Botanists are working on breeding new varieties of sorghum that don't produce the hormone, which would render the witchweed seeds powerless.

SUGARCANE

Saccharum officinarum
POACEAE (GRASS FAMILY)

There is a kind of coalesced honey called sakcharon found in reeds in India and Arabia the happy, similar in consistency to salt, and brittle enough to be broken between the teeth like salt. It is good dissolved in water for the intestines and stomach, and taken as a drink to help a painful bladder and kidneys. Rubbed on it disperses things which darken the pupils.

This odd passage from Dioscorides' five-volume book of medicine, *De Materia Medica*, described a sweet grass that had only been known in Europe since about 325 BC, when Alexander the Great brought it from India. ("Arabia the happy," by the way, refers to Yemen, which is not to be confused with "Arabia the desert" and "Arabia the stony," common terms for other parts of Saudi Arabia.) Sugarcane, and the crystallized sugar that could be extracted from it, was a novelty to the Greeks at that time, but it was well known in India and China, thanks in part to a unique anatomical advantage that allowed it to travel well.

the birth of sugarcane

Botanists believe that sugarcane was cultivated in New Guinea as early as 6000 BC. The tender young reeds were probably simply cut and chewed for a source of sweetness, but the more mature plants served another purpose: they could have been used as a building material. It's easy to see how someone could have cut a number of sturdy canes, stuck them in the ground as supports for a thatched hut, and noticed how quickly the canes put down new roots and kept growing. Sugarcane, like bamboo, turned out to be astonishingly easy to propagate. No special knowledge was required: just cut a piece, keep it damp, and put it in the ground somewhere else.

THE DRUNKEN BOTANIST

96

STEWART

Imagine, then, how easy it was to move the plant around. Sugarcane could have simply floated to Indonesia, Vietnam, Australia, and India. In fact, much early trading and cultural contact happened exactly like that. Any number of early exchanges were the result of flotsam, jetsam, and rafts blown off course. A sturdy, lightweight reed, suitable as either a building material or food, would have traveled well.

China had its own kind of sugarcane, *Saccharum sinense*. Although it was smaller than the New Guinea species, it was also tougher and able to withstand colder temperatures, poorer soils, and drought. India had its species, *S. barberi*. Some amount of cross-breeding took place between these and a few earlier, wilder species, although botanists disagree on exactly how this happened. What we do know is that the hybrids traveled well and thrived throughout warmer climates in Asia and Europe. By the fifteenth century, Europeans had a sturdy, robust, and powerfully sweet form of sugarcane to take with them on spice trading routes. The Portuguese took it to the Canary Islands and West Africa, and Columbus brought it to the Caribbean.

Once it arrived in the New World, sugarcane gave us rum, but it gave us something else, too: slavery. Starting in the early 1500s, European trading ships sailed to West Africa and went from there to sugar plantations in the Caribbean, introducing human cargo to their trading partners and opening one of the most monstrous chapters in our history. There was nothing pleasant about work in sugarcane fields. In blistering heat, the canes had to be cut by hand using enormous knives, pressed in powerful mills, and boiled in ferociously hot kettles. There were snakes and rodents and vermin of all sorts living in the fields. It was dangerous, exhausting, back-breaking work. The only way to get people to do it was to kidnap them and force them to, under penalty of death—which is exactly what happened. Slavery was abhorrent to some Europeans and early Americans: British abolitionists, for instance, refused to take sugar in their tea to protest the way in which it was manufactured. But hardly anyone refused to drink rum.

SUGARCANE CULTIVARS

Modern sugarcane varieties are given decidedly unromantic names like CP 70-1133. But some older varieties are still grown by tropical plant collectors. Many sport vivid colors, wild stripes, and far more interesting names, such as:

Asian Black	Creole	Pele's Smoke
Batavian	Georgia Red	Striped Ribbon
Bourbon	Ivory Stripes	Tanna
Cheribon	Louisiana Purple	Yellow Caledonia

the botany of a cane

Sugarcane seems, at first, like a simple plant. It's just a tall, sweet grass. But look closely and it's easy to see that quite a lot goes on inside a single stalk. The cane that emerges from the ground is segmented into joints separated by nodes. Each node holds "root primordia"—tissue that could turn into roots under the right circumstances—and a single bud, ready to grow into a stem and leaves. These highly charged little bands of tissue explain why it's so easy to propagate sugarcane. Just place a single joint with an intact node (a cutting like this is called a sett) underground, and it will unfurl "sett roots," which provide temporary nutrition, and then more permanent "shoot roots" that will anchor the plant in place and keep it alive. The bud will unfurl and become a new cane.

The stalk is made up of concentric layers, similar to a tree trunk. The outer layer, a hard, waxy rind, protects the plant from water loss. It might be yellow, in the case of a young, growing plant, or green as the chlorophyll starts to show. Red and blue anthocyanins—plant pigments that, in the case of sugarcane, protect the plant from sun damage—might turn the stalks a bright purple or burgundy red. Some varieties are even striped like candy canes.

THE DRUNKEN BOTANIST

STEWART

In the center of the stalk are soft, spongy plant tissues that transport water up from the roots and carry sugar down from the leaves. This is where the magic happens. Each joint ripens separately, which is to say that the joint closest to the ground matures until it holds the most sucrose it possibly can. The joint above it holds slightly less, and the one above that holds even less, and so on. Under ideal conditions—a long, warm season with lots of sunshine and high humidity—sugarcane elongates quickly and fills with sugar. Growers call this the grand growth period. At the end of that period, the cane is cut as close to the ground as possible, to get the joints with the highest concentrations of sugar.

If the stalk doesn't get cut, sugarcane will bloom. It produces a loose, feathery plume, sometimes called an arrow, which sits high above the leaves to catch a breeze. This is how the pollen is spread. Each plume holds thousands of tiny flowers that could each produce a single tiny seed—but on sugarcane plantations, the cane is harvested before it can reproduce, and setts are buried in the fields to start the next generation.

making rum

Hacking into a dense field of cane was difficult enough without the leaves, sharp as blades, cutting into the skin of the workers. The creatures living the fields—snakes, rats, fat fleshy centipedes and stinging hornets—delivered another round of unwelcome surprises. One solution was to set fire to the fields before the harvest, driving the vermin out and clearing most of the vegetation away. This is still done in some cane fields today, even on modern farms that use heavy equipment to do the harvesting.

DAIQUIRI

1½ ounces **white rum**
1 ounce **simple syrup**
¾ ounce **fresh-squeezed lime juice**

A classic daiquiri is made with nothing but these three ingredients. Shake over ice and strain into a cocktail glass.

A SUGAR PRIMER

SUGAR—OR THE SIMPLE SYRUP MADE BY HEATING
EQUAL PARTS SUGAR AND WATER—IS A VITAL COCKTAIL
INGREDIENT. BUT THERE ARE MANY KINDS OF SUGAR,
SOME BETTER SUITED TO DRINKS THAN OTHERS.

BROWN SUGAR is refined sugar with molasses sprayed on for flavor
and color.

DEMERARA OR MUSCOVADO SUGAR are two forms of large-
grained raw sugar with some molasses coating or residue.

POWDERED SUGAR contains a small amount of cornstarch or flour
to prevent clumping, which is nice in baked goods but will gum up a
drink. Avoid it in cocktails.

SUPERFINE SUGAR (also called baker's sugar or caster sugar) is
ordinary, granulated sugar that has been finely ground so that it
dissolves quickly. Ideal for cocktails.

TURBINADO OR RAW SUGAR is made from the first extraction of
cane juice. The granules tend to be larger and contain some molasses
flavor. Makes a richer simple syrup, although it may take longer
to cook.

Once the cane is harvested, it is highly perishable and must get to the mill quickly before bacteria start eating the sucrose and robbing the sugar factory of its product. So as soon as it is cut, the cane is chopped, crushed, and milled to extract the juice. On the French Caribbean island of Martinique this fresh juice is fermented and distilled directly to make rhum agricole; in Brazil the fresh cane juice becomes *cachaça* (pronounced "cachasa.") But most of what we know as rum comes from molasses, not cane juice.

BAGASSE: THE SUGARCANE RESIDUE LEFT OVER AFTER THE JUICE IS PRESSED FROM THE STALK. USED FOR FUEL, LIVESTOCK FEED, BUILDING MATERIALS, AND COMPOSTABLE PACKAGING.

When sugar is processed, the juice is filtered, purified, and heated to crystalize the sugar. Left behind is a dark, rich syrup: molasses. If it's going to be made into rum, the molasses is then fermented with water and yeast to make a wash of 5 to 9 percent alcohol. That is then distilled, originally in simple pot stills, and now in more sophisticated column stills.

On plantations, rum was a cheap drink for the workers, not a fine export. The owners of those farms probably drank port or brandy, not rum. The first colonists arriving in New England, lacking a quick and easy way to make alcohol, imported molasses from the Caribbean to turn into rum. But that was an act of desperation and, later, an act of defiance. The Molasses Act of 1733 was an attempt by the British to force the colonies to buy British, not French, molasses, by imposing heavy import taxes on French products. Such laws only fueled the colonists' outrage and kindled the American Revolution. John Adams, writing to his friend William Tudor in 1818, said, "I know not why we should blush to confess that molasses was an essential ingredient in American independence. Many great events have proceeded from much smaller causes."

Sugarcane cultivation became an American enterprise as well. It grows on nine hundred thousand acres in Florida, Louisiana, Texas, and Hawaii. But most rum still comes from the Caribbean. The

A FIELD GUIDE TO SUGARCANE SPIRITS

AGUARDIENTE: A generic Spanish term for clear neutral spirits or brandy; in many Latin American countries it refers to a sugarcane-based spirit.

BATAVIA ARRACK: A high-proof (50 percent ABV) Indonesian spirit distilled from sugarcane and fermented red rice. A key ingredient in classic punch recipes.

CACHAÇA: The main ingredient in a Caipirinha, this Brazilian spirit is distilled from fresh sugarcane juice. (The other ingredients are sugar and lime juice.)

CHARANDA: A Mexican spirit, often called "Mexican rum."

LAKANG HARI IMPERIAL BASI: A sugarcane-based wine from the Philippines.

PUNCH AU RHUM: French liqueur made with a rum base.

RHUM AGRICOLE: Rum from the French West Indies distilled from sugar-cane juice, not molasses.

RUM: Alcohol distilled from fermented sugarcane juice, syrup, molasses, or other sugarcane by-products, distilled to less than 80 percent ABV and bottled at or below 40 percent ABV.

RUM-VERSCHNITT: A German mixture of rum and other alcohol.

SUGARCANE OR MOLASSES SPIRIT OR VODKA: A generic term for a clear, neutral, high-proof spirit distilled from sugarcane.

VELVET FALERNUM: Rum-based sweet liqueur that is flavored with limes, almonds, cloves, and other spices, a key ingredient in tropical rum-based drinks like the mai tai.

reason for this is, in part, an accident of history—the oldest and best-known distilleries are by necessity located where the sugarcane grew first. It is also an accident of climate. When rum goes into a barrel, the same wonderful interplay of alcohol and wood that makes whiskey so mellow and smooth also happens with rum. But in the tropics, it happens much, much faster. A barrel of rum (often a used bourbon barrel) loses a whopping 7 to 8 percent of its alcohol per year as the wood expands and softens in the steamy heat. What might take twelve years to accomplish in Scotland happens in just a few years in Cuba. For this reason, dark, well-aged Caribbean rums are astonishingly rich and complex after just a short repose in wood.

the naval spirit

Although rum is a drink of the Americas, its history is inextricably tied with that of the British navy, and a surprising number of recipes, colloquialisms, and strange bits of technology came out of the navy's long relationship with its favorite spirit.

In the 1500s, sailors were given beer to drink, in part to keep them happy and in part because water, without any alcohol to kill bacteria, spoiled quickly at sea. But even beer went bad on long voyages, so rum became the ration of choice. Giving sailors an entire pint of rum turned out to be a bad idea—they'd drink the whole thing and ignore their duties—so the solution was to mix it with water, lime juice, and sugar, which improved the taste and combated scurvy. This grog (it was not strong enough to call a daiquiri, even though the ingredients were more or less the same) could be doled out twice a day without imperiling the ship.

It's easy to see how disgruntled sailors might start to wonder if their rum had been diluted a little too much. They demanded proof that they were getting the rum they were entitled to. There were no hydrometers in those days (a hydrometer is an instrument that measures the density of a liquid as compared to water, thereby measuring alcohol content), so a method was developed using a material ships always had on board: gunpowder. A quantity of gunpowder, mixed with rum, would not ignite if the rum was watered down. It would have to contain about 57 percent alcohol to catch on fire. In the presence of the crew, the ship's purser would mix the rum and gunpowder and light it on fire, offering "proof" of its potency.

MOJITO Y MAS

3 sprigs fresh spearmint
¾ ounce fresh-squeezed lime juice
1 ounce simple syrup
1½ ounces white rum
Soda water
For the variation: Sparkling wine (a dry Spanish *cava* works well)
  and fresh fruit

In a cocktail shaker, muddle 2 spearmint sprigs, the lime juice, and the simple syrup. Add the rum, shake with ice, and strain into a highball glass filled with crushed ice. Top with soda water and garnish with the remaining sprig of spearmint.

VARIATION: *The Mojito y Mas makes use of any garden-fresh, seasonal fruit available. Peaches, plums, apricots, raspberries, and strawberries work best. Make the recipe as usual, but fill the highball glass with a mixture of crushed ice and chopped fruit. Add the rum, top with sparkling wine instead of soda (a dry Spanish* cava *works well), and go sit in the sun.*

British proof is still based on this standard: a bottle is 100 proof if it contains 57 percent alcohol. In the United States, the math is easier: 100 proof is the equivalent of 50 percent alcohol.

In 1970, the British navy discontinued its rum rations. Sailors protested, wearing black armbands and appealing to Prince Philip, a retired navy man himself, to "save our tot." But it was no use. Eliminating the rations saved money and helped make sure that the sailors steering submarines were at least as sober as civilians driving cars. The tradition has been gone for over forty years, but some rum distillers continue to offer a "navy strength" version bottled at 57 percent alcohol.

⊰ SUGAR BEET ⊱

Beta vulgaris
CHENOPODIACEAE (GOOSEFOOT FAMILY)

In 1806, Napoleon Bonaparte found himself in a bit of a bind. He'd issued an order, known as the Berlin Decree, banning the importation of any British goods. That meant no tea, no warm British wool, no indigo dye, and no sugar for the people of France. At that time, most sugarcane production in the Caribbean was under British control. Knowing that this would be a disaster for Parisian pastry chefs, Napoleon hatched a plan to refine sugar from beets.

He turned to botanist Benjamin Delessert to develop a method. Soon there were six experimental stations operating around France, with a hundred students learning the process. Farmers were required to plant thousands of acres in beets. Forty factories were pumping out over three million pounds of sugar. In 1811, Napoleon wrote that the British could throw their sugarcane into the Thames, because Europe would have no further use for it. But after his exile, the political winds shifted again, and sugarcane came back to France.

The modern sugar beet is a hefty, white variety grown for its high sucrose content—18 percent, which is higher than most sugarcane. It can grow to a foot long and weigh five pounds. A close relative of chard and amaranth, the beet is probably native to the Mediterranean, where it emerged as a more domesticated form of the wild sea beet, *Beta vulgaris* subsp. *maritima*, also called wild spinach. Although botanists developed a method for boiling it to make a sweet syrup in the late sixteenth century, it wasn't used as a sweetener until varieties had been bred that contained higher sugar levels. That breakthrough, along with technological advances and sheer necessity, finally made it possible to extract a reasonable amount of sugar from a beet.

Today a quarter of the world's sugar supply comes from beets, with the United States, Poland, Russia, Germany, France, and Turkey leading the way. Fifty-five percent of the sugar produced in the United States comes from sugar beets, mostly grown in the upper Midwest and western states. America consumes all the sugar it grows and imports more, mostly from Latin America and the Caribbean, to satisfy its sweet tooth.

The process is similar to that of sugarcane. The juice is extracted with hot water rather than a mill, but after that, it is filtered, heated, and the sugar crystals are separated from the molasses. The sugar extracted from beets is identical to that of sugarcane, but the molasses is different: beet molasses is bitter and unpalatable because of the non-sugar residues left behind. It can be fed to livestock and has even been sprayed on icy roads to help the salt stick.

Of interest to drinkers, however, is the fact that beet molasses is sold to commercial producers of yeast, who mix it with cane molasses to provide a sugary medium for large-scale yeast cultivation. After being raised on molasses, the yeast is filtered, compressed, and sent out to breweries, distilleries, and bakeries. So in a way, all alcohol begins with beet sugar.

Some spirits are produced from beet sugar, although it may not be obvious: so-called rectified spirits or neutral spirits can be made from beet sugar and used as a base in liqueurs or to adjust the proof of a spirit. Orange liqueurs like triple sec, and many brands of absinthe and pastis, are made with a beet sugar alcohol base. And around the world, a few rums are made from beet sugar, including the Swedish Altissima and the Austrian Stroh 80. Craft distilleries in the United States have attempted it as well. Michigan's Northern United Brewing Company makes a version of rum based on beet sugar, and Wisconsin's Old Sugar Distillery distills an anise-flavored ouzo and a honey liqueur from beet sugar.

WHEAT

Triticum aestivum
POACEAE (GRASS FAMILY)

As one of the oldest cereal grains, wheat would seem like the logical candidate for the title of most ancient and primary beer ingredient. It was domesticated over ten thousand years ago in the Middle East and arrived in China by 3000 BC. As a food source, it has everything going for it: protein, flavor, durability, and a wonderful elasticity that allows bread to rise. But some of the very qualities that make it good to eat also make it difficult to ferment. In fact, brewers and distillers consider it one of the trickier ingredients to work with.

To understand this problem, think about it from the plant's per-spective. A grain of any kind is, of course, a seed; it represents the plant's next generation, its shot at immortality. To ensure the seed's success, the plant stores sugar next to the embryo in the form of starch. But sugar alone is not enough: a seedling needs protein, too. So embedded within the starch is a matrix of protein. When the seed drops to the ground and gets a little damp, enzymes go to work busting apart that starch so that the seedling will have some sugar to eat. But first, they have to get past the protein.

Wheat is particularly good at taking up nitrogen, one of the build-ing blocks of protein. Those wheat proteins are fairly flexible in the sense that they'll take extra nitrogen when they can get it. That makes for a strong matrix wrapped around the starch. This is good news for bakers: a healthy amount of protein makes a great loaf of bread. Those wheat proteins come together in the presence of water to form gluten—that sticky, stretchy stuff so important in dough.

That's why farmers, for thousands of years, have selected strains of wheat that are high in protein and eager to take up nitrogen. Their selection process, however, is not as beneficial for brewers.

BUCKWHEAT

BUCKWHEAT (*FAGOPYRUM ESCULENTUM*) IS NOT WHEAT AT ALL, but a flowering plant in the knotweed family. It is closely related to dock and sorrel, two wild European herbs. The dark, triangular seeds are enclosed in a hull that makes up about a fourth of the seed's bulk. When the hull is removed, what's left is called a buckwheat groat.

In addition to its use as flour in pancakes and noodles, and in cereal (such as European kasha), buckwheat is used in Japan to make the spirit *shochu*, and it's turning up in vodkas and in beers as a gluten-free alternative. The French Distillerie des Menhirs makes what it claims is the world's only buckwheat whiskey, called Eddu Silver.

In a brewer's mash, the starch is so tightly bound up in the protein matrix that some of it is inaccessible. The equation is simple: more nitrogen means more protein, which means less sugar and therefore less alcohol. Complicating matters is the fact that wheat can get gummy in a mash tub, and the stray bits of protein left behind after fermentation make the brew cloudy.

a touch of wheat

So even though wheat went into ancient brews, it was never used alone. Egyptians mixed their wheat with barley, sorghum, and millet to come up with a more workable recipe. A fine wheat beer tradition developed in Germany, starting in the Middle Ages, but even those beers consisted of only about 55 percent wheat, with barley making up the rest of the grain. Russian distillers made early vodkas from a mixture of wheat, barley, and rye, and Scotch and Irish whiskey makers perfected the art of making whiskey from a similar blend, with a little corn added. Without the help of those other grains, wheat would not make much of a drink.

Why bother with wheat at all, if it is so difficult to work with? Try a German *Hefeweizen* and you'll have your answer. There's a definite bread and biscuit aroma that is impossible not to love. Wheat is also smooth and round and easygoing. It settles in happily with the other flavors around it. German wheat beers are known for their spicy, citrusy character; this comes not so much from the hops but from special strains of yeast that go to work on those wheat sugars and produce their own unique flavors. Those beers are also known for their thick, foamy head, which is mostly dissolved wheat protein. Many brewers add a touch of wheat to their grain mixture just for the foam.

In vodka and whiskey, wheat makes for a light, smooth spirit, and this can be a wonderful thing. Any number of bourbon drinkers will say that they've tasted all kinds of fancy bourbons but keep returning to Maker's Mark. Why? It's the wheat. Most bourbon contains a little rye in addition to corn and barley, but Maker's uses wheat instead of rye. That smooth, sweet flavor, quite different from the spicy bite of rye, explains why Maker's is such a crowd-pleaser. The influence of wheat is even more obvious in some of the new American "straight wheat" whiskies, in which wheat makes up at least 51 percent of the blend. And the sheer palatability of wheat is still more evident in three of the world's most popular vodkas: Grey Goose, Ketel One, and Absolut.

WHAT ABOUT *the* LEMON WEDGE?

Wheat beers are often served with a lemon wedge to highlight their natural citrus flavors, but some beer aficionados consider this a sacrilege. They argue that good beer should never require additional flavorings. In certain company, what one does with a lemon wedge and a glass of wheat beer could make or break a friendship. It's your drink, so do what you want—but proceed with caution.

Until recently, the needs of brewers and distillers were ignored by wheat farmers. Planting a hard, high-protein wheat, and feeding it plenty of nitrogen, is a good strategy for a farmer who wants to feed the world. But if the farmer wants a nice glass of whiskey at the end of the day, putting in a few fields of soft wheat would help. Wheat varieties are distinguished by growing season (winter vs. spring), by color (amber, red, or white), and by protein content, with soft wheat being lower in protein. Now plant breeders are more focused on breeding low-protein varieties, and farmers who want to grow for brewers are using less nitrogen fertilizer in their fields. Wheat for brewing and distilling might make up only a small fraction of the 689 million tons produced around the world, but without it, beer, whiskey, and vodka wouldn't be the same.

DRINK YOUR WHEAT

There are thousands of wheat varieties around the world. Most of the wheat sold to brewers and distillers is simply labeled by type, such as "soft red winter wheat." But here are a few of the specific varieties you might find in your bottle.

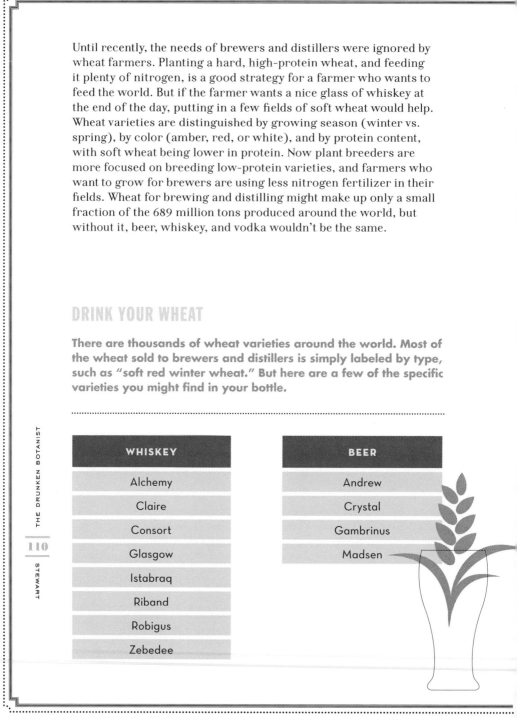

WHISKEY
Alchemy
Claire
Consort
Glasgow
Istabraq
Riband
Robigus
Zebedee

BEER
Andrew
Crystal
Gambrinus
Madsen

STRANGE BREWS

A STIFF DRINK CAN BE MADE
FROM MORE THAN BARLEY AND GRAPES.
SOME OF THE MOST EXTRAORDINARY
AND OBSCURE PLANTS HAVE BEEN
FERMENTED AND DISTILLED.
A FEW OF THESE ARE DANGEROUS,
some are downright bizarre,
AND ONE IS AS ANCIENT AS DINOSAURS—
BUT EACH REPRESENTS A UNIQUE
CULTURAL CONTRIBUTION TO OUR GLOBAL
DRINKING TRADITIONS.

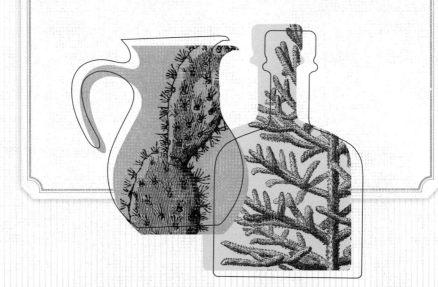

BANANA

Musa acuminata

MUSACEAE (BANANA FAMILY)

The banana tree is actually not a tree but an enormous perennial herb. It is disqualified from being a tree because its stem contains no woody tissue. Most of us have only ever eaten one kind of banana, the Cavendish, which is what our supermarkets carry, but there are in fact hundreds of cultivars, including the so-called beer bananas of Uganda and Rwanda. Farmers prefer to grow beer bananas (as opposed to cooking bananas, also known as plantains), because they can process the fruit into a highly profitable beer that, while short-lived, does not perish as quickly as the bananas themselves do. Transformed into beer, the bananas are easier to get to market.

The traditional method is to pile ripe, unpeeled bananas into a pit or basket. People tread on them to extract the juice, much like the stomping of grapes. The juice is roughly filtered through grass and left to ferment in a gourd, to which sorghum flour might be added. After a couple of days, the cloudy, sweet and sour beer is ready to drink. It can be bottled and stored for two or three days at the most.

While Ugandan banana beer is usually a homemade affair, brewers have made commercial versions. Chapeau Banana is a Belgian lambic. The British Wells & Young's Brewing Company makes Wells Banana Bread Beer, and the Mongozo brewery in the Netherlands offers a banana beer made in the African style with fair-trade bananas.

CASHEW APPLE

Anacardium occidentale

ANACARDIACEAE (CASHEW FAMILY)

Most people have never taken a cashew nut out of its shell. There's a good reason for this: the cashew tree is a close relative to poison ivy, poison oak, and poison sumac. Like its cousins, it excretes a nasty, rash-inducing oil called urushiol. The shells have to be carefully steamed open to extract the edible, urushiol-free nut inside.

The nut hangs from a small fruit called the cashew apple. (In botanical terms, the cashew apple is actually a pseudo-fruit because it does not contain any seed; the real fruit is the cashew nut hanging below it.) This fruit, which is also free of the noxious oil, is used in India to make a fermented drink called *feni.*

The cashew tree, native to Brazil, was described in 1558 by French botanist André Thevet. In a woodcut, he depicted people squeezing the fruit while it still hung on the tree. Portuguese explorers brought the cashew to their colony in Mozambique and to the eastern coast of India. European tastes in liquor called for new uses of the cashew: in 1838, a report on the drinking habits of people in the West Indies included a description of a punch, presumably rum-based, flavored with the juice of the cashew apple.

These squat, fast-growing trees, which climb to about forty feet tall and stretch twice as wide, were planted in India with the idea that they would help with erosion control. Cashew trees are now also found in East Africa and throughout Central and South America, but the world's supply of cashew nuts comes primarily from Brazil and India.

Cashew apple *feni* (sometimes *fenny* or *fenni*) is still made in the tiny Indian state of Goa, which was occupied by Portugal from 1510 through 1961. It's a popular vacation spot for European tourists, who seek out the local beverage while on holiday.

The apples signal their ripeness by dropping from the tree or separating with only the slightest pressure; they must then be crushed immediately because they spoil quickly. To make *feni,* locals separate the cashew apple, which they call *caju,* from the nut. The fruit is placed into a pit and stomped, sometimes by children wearing rubber boots. The juice is set aside to make a lightly fermented summer drink called urak. Some of this fermented beverage is then distilled in a copper pot to about 40 percent ABV; it is this strong, clear drink that is called *feni.* The locals enjoy it with lemonade, soda, or tonic water.

CASSAVA

Manihot esculenta

EUPHORBIACEAE (SPURGE FAMILY)

The cassava root has been an important food source for people in impoverished and famine-prone areas around the world. Even today it feeds four hundred million people in Africa, Asia, and Latin America. The starchy roots, which grow to over three feet in length and weigh several pounds, do offer some nutrition, particularly vitamin C and calcium, but they are also poisonous if not processed properly. In order to leach the cyanide out of the roots, they must be soaked in water, cooked, or pounded into a flour and spread out on the ground for several hours to allow the cyanide to break down or escape into the air. So-called sweet varieties require less processing than the more nutritious, but also more poisonous, bitter varieties. Neither is necessarily safe to eat raw.

In spite of these difficulties, the cassava—also called manioc root—is a staple food because it is drought-tolerant and fairly easy to grow. In the Caribbean and parts of Latin America, especially in Brazil, Ecuador, and Peru, manioc beer (called *ouicöu* on the islands) is made by peeling and chopping the root, boiling it in water, and then chewing the pulp and spitting it back into the mash. This introduces amylase, an enzyme in saliva that helps convert starch to sugar. It is then brought back to a boil, and sugar, honey, or fruit might be added to help increase the alcohol content and improve the flavor.

Cassava is native to South America; it was domesticated in Brazil by about 5000 BC. Although it was introduced to East Africa by the Portuguese in 1736, it wasn't widely grown there until the twentieth century. Any cassava beer-making tradition in Africa is therefore relatively recent. Multinational beer conglomerate SABMiller, known for such brands as Coors Light and Henry Weinhard's, recently announced plans to brew cassava beer in Angola, sourcing ingredients from local farmers and selling the beer at a lower cost, in hopes of creating not only jobs but also a new market for beer among thirsty, impoverished Africans.

BUGS *in* BOOZE: HONEYBEES

--- *Apis* spp. ---

No insect is more important to the history of alcohol than the honeybee. Just about every kind of fermentable fruit—from grapes to apples to the strange and lovely tamarind—is pollinated by bees, which means that without them, we risk a sudden and shocking sobriety, not to mention scurvy and starvation. But there is a more direct route from bees to intoxication—honey.

Even before the advent of beekeeping in Egyptian times, honey was collected in the wild. Primitive drawings of bee hunters climbing cliffs to rob hives of honey date back to the Neolithic and Mesolithic eras. The earliest beehives, called skeeps, were made of simple baskets that could at least be hung in a more convenient location, making long treks through the forest in search of honey unnecessary.

The earliest form of honey wine, or mead, probably came about when honeycomb was drained of most of its honey and then soaked in water to remove the rest. This honey water would have fermented naturally in the presence of wild yeast. Later, when beekeepers realized that they could get lighter, sweeter honey by placing bee-hives near particular crops like clover, alfalfa, and citrus, the wild honey collected in forests went first to mead, while more refined, cultivated honey was preferred as a sweetener.

The Greeks used the word *kykeon*, meaning "mixture," to refer to a strange beverage that combined beer, wine, and mead. In Homer's *Odyssey*, Odysseus's crew was drugged with *kykeon* by Circe, a sorceress who turned them into pigs. Greek and Roman mead-making traditions spread throughout Europe, but Africans had their own methods as well. The Azande tribe in north-central Africa made mead, and in Ethiopia, a kind of mead called *tej,* or *t'edj,* is still widespread. The recipe calls for about six parts water to one part honey. After a few weeks of fermentation, usually in a pottery or gourd vessel, the drink achieves the alcohol content of wine and is ready to drink. It is sometimes flavored with the bitter leaves of a buckthorn shrub (*Rhamnus prinoides*), or with khat (*Catha edulis*), a leaf that is chewed as a mild stimulant. And in sub-Saharan Africa, tamarind or other fruit would be added to the mixture of honey and water to create an even sweeter drink.

In Paraguay, the Abipón tribe simply mixed honey and water and waited a few hours, which produced a mildly alcoholic beverage fermented on wild yeast. The Sirionó of Bolivia added honey to a gruel of corn, manioc root, or sweet potato, which fermented for a few days until it became as strong as beer. Even early Americans made their own kind of mead, a murky, dark concoction that settlers said was so strong it made them hear the bees buzz.

Today's high-quality meads have a bright, floral flavor that is some-times enhanced with fruit, herbs, or hops to change the character of the drink. A beer and mead hybrid called braggot is made by some craft brewers; Beowulf Braggot from Dogfish Head Brewery is one such example. Although mead can be distilled into a stron-ger spirit (sometimes called honeyjack), it's not commonly made. Hidden Marsh, a distillery in Seneca Falls, New York, makes its Bee Vodka from honey; it is surprisingly smooth, with only the slightest hint of sweetness. But the best true honey flavor can be found in a German liqueur called Bärenjäger, which even comes in a bottle with a beehive-shaped cap.

DATE PALM

Phoenix dactylifera

ARECACEAE (PALM FAMILY)

I n 2005, an archeologist in Israel had a simple but stunning idea: why not try to germinate the two-thousand-year-old date palm seeds that had been sitting in storage? While old seeds from archeological excavations had been sprouted before, nothing this old had ever been resurrected. But date palms produce what botanists call orthodox seeds, which means that they remain viable long after they have thoroughly dried. (The opposite of an orthodox seed is a recalcitrant seed, which can only be sprouted while fresh and damp. Avocados, for instance, produce recalcitrant seeds.)

This particular ancient seed came from an excavation at Masada, in Israel, where Jewish Zealots committed mass suicide in 73 AD rather than submit to Roman rule. The seed had been found at that site and stored carefully away until the day archeologists decided to sprout it. If plants could act surprised, this one certainly would have been startled to awaken, after a nearly two-thousand-year slumber, in a modern greenhouse, housed in a plastic pot and fed by drip irrigation. This particular variety of palm, called a Judean date palm, went extinct around 500 AD, making it even more astonishing that the plant was resurrected from the dead. Its caretakers are still waiting to find out if they have sprouted a boy or a girl; they hope for a girl so they can sample a long-vanished fruit.

Date palm fruit is a staple of Mediterranean, Arabic, and African cuisine. Date palm wine, however, comes not from the fruit but from the sugary sap of the tree. This is an ancient drink, depicted in Egyptian paintings dating back to at least 2000 BC. The process to make it has not changed much over the millennia. To get the sap flowing, the tree is tapped, usually by inflorescence decapitation, which is the technical term for cutting off a flower. In some cultures, an elaborate ritual of bending, twisting, beating, kicking, and otherwise abusing the flower precedes its decapitation. All of this leads to a more productive flow of sap.

Other palm tree species, including the coconut tree (*Cocos nucifera*), are tapped for their sap throughout Asia, India, and Africa, and for each tree there is a different technique. Sometimes the tree is cut down entirely. Sometimes a hole is cut in the trunk at the very top of the tree, bringing it to the verge of death and sometimes killing it. And in many cases, the tree is simply scraped or punctured as a maple tree would be.

Once the sap is collected, it can be used as a sweetener or cooked into a block of sugar called jaggery. If left alone, it begins to ferment almost immediately, thanks to wild yeast in the air and on the gourds used to collect it. Within hours, a sweet, mild, alcoholic beverage is ready to drink. The fermentation can continue for a few more days, which allows the alcohol content to rise slightly, but the yeast eventually give way to bacteria—and a bacterial fermentation yields vinegar, not wine. At some point during the fermentation the drink reaches the perfect balance of alcohol, sweetness, and a mild acidity, and at that moment it must be consumed at once. Don't go looking for date palm wine in a liquor store; it won't keep long enough to bottle it. The wine can also be distilled into a stronger spirit, sometimes called arrack, which is a general term referring to spirits made from sugary sap.

In West Africa alone, over ten million people enjoy date palm wine—but unfortunately, humans aren't the only ones who love it. In Bangladesh and India, fruit bats visit the gourds and drink the fresh sap that is collecting there. The bats carry a serious disease called Nipah virus, which they can leave behind in the date palm sap. This has been responsible for transmitting the virus from bats to humans. The solution? Health-care workers are scrambling to find a way to tap the date palm without letting the bats have a sip.

JACKFRUIT

Artocarpus heterophyllus

MORACEAE (MULBERRY FAMILY)

The jackfruit may be the largest fruit from which an alcoholic beverage is made. It grows to three feet in length and can weigh as much as a hundred pounds. The fruit's strange, rubbery exterior is covered in spiky, conelike structures, each of which represents a spent flower. Inside, there is a seed for every flower that once bloomed on its surface: as many as five hundred seeds can come from a single jackfruit. When ripe, the fruit emits a foul odor from the rind, but the flesh is mild and sweet. It flavors desserts, curries, and chutneys.

The tree, a close relative of breadfruit, grows throughout India, as well as parts of Asia, Africa, and Australia. In India, wine is made by soaking the pulp in water, sometimes with extra sugar, and allowing it to ferment naturally for up to a week, at which point the alcohol content reaches 7 to 8 percent and the drink becomes mildly acidic, but still light and fruity.

MARULA

Sclerocarya birrea subsp. *caffra*
ANACARDIACEAE (CASHEW FAMILY)

The marula tree, a close relative of mango, cashew, poison ivy, and poison oak, is native to Africa. Its yellowish white fruit, about the size of a plum, has a flavor similar to lychee or guava. Because it is unusually high in vitamin C, the fruit is an important part of the traditional diet in southern and western African nations. It can be made into wine, also called marula beer, by soaking the fruit in water and allowing it to ferment. It is also distilled and blended with cream to make Amarula Cream, a dessert drink that tastes very much like an Irish cream liqueur.

Because the tree has, since at least 10,000 BC, served so many purposes in traditional African culture—for food, medicine, rope fiber, wood, cattle feed, oil, and resin—efforts are under way to protect and conserve the marula. South Africa's spirits manufacturer Distell purchases the fruit from local pickers, providing them a source of income and donating money to community projects. Development experts believe that, with good oversight, the global trade in Amarula Cream can provide an economic incentive to preserve the trees while helping impoverished families.

The elephant on the bottle of Amarula Cream reminds drinkers of a popular, but widely discredited, story about the marula: that elephants can get drunk by slurping up overripe, fermenting fruit that has fallen from the tree. Drunken elephant stories started circulating around 1839 and continue today, with Internet videos purporting to show intoxicated elephants stumbling around.

But scientists have proven otherwise. Elephants do not pick rotten fruit off the ground; instead, they very deliberately select ripe fruit from the tree. There is also the sheer difficulty in getting an elephant drunk. It would take about a half gallon of pure alcohol, which would require the elephant to rapidly consume some fourteen hundred rotten marula fruits—something no elephant has ever been interested in attempting.

MONKEY PUZZLE

Araucaria araucana

ARAUCARIACEAE (ARAUCARIA FAMILY)

The poet Marianne Moore called it "a conifer contrived in imitation of the glyptic work of jade / and hard-stone cutters, / a true curio in this bypath of curio-collecting." And in fact, the monkey puzzle tree is a curio, an oddity of the sort prized by Victorian plant collectors. It is also quite likely the oldest plant in the world from which an alcoholic beverage is made.

Monkey puzzle trees originate in Chile and Argentina. Their ancestry goes back at least 180 million years, placing them squarely in the middle of the Jurassic period. The trees themselves are vaguely reptilian: the tough, diamond-shaped leaves arranged in tight geometric whorls bring to mind the scales of a lizard. Stand back for a better view and the tree cuts an odd figure in the landscape. From a single trunk emerge wildly curving branches that give it the madcap appearance of a tree from a Dr. Seuss drawing.

A Scottish surgeon and naturalist named Archibald Menzies traveled the world as a ship's doctor in the late eighteenth century; on one of these voyages, he was served the nuts of the monkey puzzle tree. He managed to save a few seeds and got them to grow, setting off a monkey puzzle craze in the United Kingdom. One of them lived at Kew Gardens for almost a century. There are no monkeys in the tree's homeland; the name was given to it by the English, who thought that even their so-called poor relations—monkeys—would have a difficult time climbing it.

The tree reaches over 150 feet in height and can live to be a thousand years old. A monkey puzzle takes twenty years to reach sexual maturity and is dioecious. The pollen travels from males to females on the wind, and once pollinated, the seed cones take two years to mature. By the time they do fall off the tree, they are the size of coconuts and contain about two hundred seeds, each larger than an almond.

In the wild, rats and parakeets pick up the seeds and disperse them from the mother plant. But if there are people around—particularly the Pehuenche people who inhabit the tree's native range in the Andes—the seeds will quickly be gathered up. They can be eaten raw or roasted, pounded into flour to make bread, or brewed into a mildly alcoholic ceremonial drink called *mudai*. To make *mudai*, the seeds are boiled and allowed to ferment naturally for a few days; to speed things up, they can be chewed and spit back into the mixture, which adds enzymes from the saliva to break down the starches. Once the mixture has stopped bubbling, it is poured into special wooden bowls or jars for the festivities.

The Chilean government has declared the monkey puzzle tree a national monument, making *mudai* quite possibly the world's only alcoholic beverage derived from a national monument.

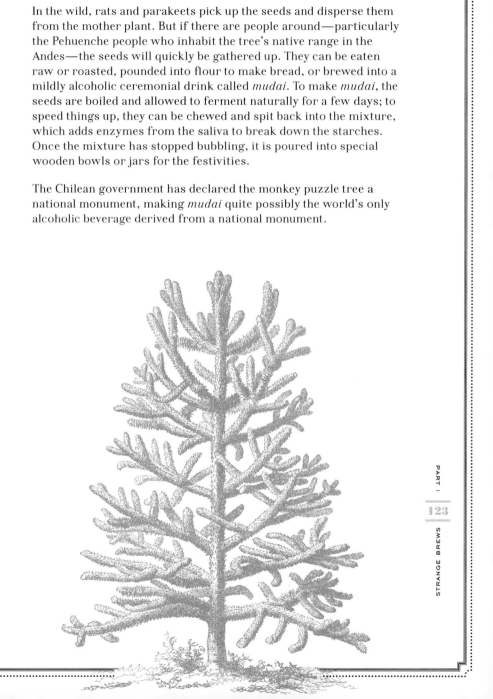

✣ PARSNIP ✣

Pastinaca sativa
APIACEAE (CARROT FAMILY)

If barley be wanting to make into malt,

We must be content and think it no fault,

For we can make liquor to sweeten our lips,

Of pumpkins, and parsnips, and walnut tree chips.

—EDWARD JOHNSON, 1630

This historic ditty shows that colonists arriving in the New World were willing to try anything to get their hands on a drink—even if that meant turning parsnips into wine. Parsnips are a carrot relative native to the Mediterranean; they've been a staple food since at least Roman times. Before potatoes, a New World crop, were introduced to Europe, parsnips were the starchy, nutritious, winter root vegetable people turned to for a satisfying meal. It's no wonder colonists made it a priority to plant parsnips when they arrived in New England.

And surely the colonists were thinking not just of mashed parsnips and butter for their winter meal. They were also thinking of parsnip wine, a fine old English tradition. This is one of many "country wines" that were popular in rural England and throughout Europe. Anything with a little sugar or starch—from gooseberries to rhubarb to parsnips—was fair game for a home brew.

Traditional parsnip wine was made by boiling parsnips to soften them and then combining them with sugar and water. Wild yeast would start the fermentation. The wine was then stored for six months to a year before drinking. It was light, sweet, and clear, although the best *Cassell's Dictionary of Cookery* could say about it in 1883 was that it was "highly spoken of by those who are accustomed to home-made wines."

WARNING: DON'T TOUCH THAT

Wild parsnip is a weed across much of North America and in Europe. Its foliage can cause serious, blistering rashes. The domesticated varieties taste better, but the leaves can still irritate, so always wear gloves around parsnip.

PRICKLY PEAR CACTUS

Opuntia spp.

CACTACEAE (CACTUS FAMILY)

The fruit of the prickly pear cactus, called the *tuna* in Mexico, is not easy to eat. The sharp prickles, or glochids, have to be scraped, burned, or boiled away. The flesh is then either scooped out of the peel and eaten fresh, or pressed into juice. All this effort is entirely worthwhile: the fruit has been an important source of vitamins and antioxidants for centuries. It has also been fermented into wine. The Chichimeca people of central Mexico, for example, traveled to follow the bloom cycle of the cactus and make their seasonal brew.

Spanish explorers and missionaries realized that the prickly pear cactus was an important food source in the desert, and not just because of its fruit. The fleshy green pads could also be peeled, cut into strips, and eaten as a vegetable, called nopales. Soon the cacti were planted around missions and transported back to Spain, and from there they went around the world.

The prickly pear cactus was once grouped together with the cholla, but botanists have recently separated them, putting prickly pears in their own genus. About twenty-five species of Opuntia have been identified, and some species, such as *Opuntia humifusa*, grow not just in the desert but throughout most of the eastern United States.

Prickly pear juice, syrup, and jam are now widely available, and prickly pear mojitos and margaritas show up on cocktail menus throughout the Southwest. Distillers are working with it as well: a prickly pear liqueur called Bajtra is made in Malta; a distiller on St. Helena distills it into a spirit called Tungi; a prickly pear vodka comes from Arizona; and Voodoo Tiki sells a prickly-pear-infused tequila.

PRICKLY PEAR SYRUP

If you're lucky enough to have access to fresh prickly pears, make a batch of syrup and keep it in the freezer. (Otherwise, prickly pear juice and syrup are available from specialty food retailers.) You can add a dollop to sparkling wine, mix it into a margarita recipe, or experiment with any cocktail that calls for fruit and sugar.

10 to 12 prickly pear fruits
1 cup water
1 cup sugar
1 ounce vodka (optional)

Prickly pear fruits sold in markets generally have the spines removed. If you've harvested the fruit yourself, handle it with metal tongs, as gloves may not protect you. Use a vegetable scrubber to brush the spines off. Then cut off both ends of the fruit, and make one cut from top to bottom. The skin can then be easily sliced away.

Chop the fruit roughly, combine with water and sugar, and bring to a boil. Use a strainer to separate the seeds and pulp from the syrup. Store the syrup in a glass jar in the freezer. Adding a little vodka keeps the syrup from freezing solid without significantly changing the character of the drink you make with it.

PRICKLY PEAR SANGRIA

Thin slices of fruit: lemon, lime, orange, prickly pear, mango,
    apple, and so on
4 ounces brandy or vodka
2 ounces triple sec or another orange liqueur
1 bottle dry white wine, such as a white Spanish Rioja
2 ounces prickly pear syrup (see p. 127)
A 6-ounce split of Spanish *cava* or other sparkling wine (optional)

*Soak the fruit in the brandy and triple sec for at least 4 hours. Combine
the wine and prickly pear syrup in a glass pitcher; stir vigorously
and add more syrup if you'd like to deepen the color. Stir in the fruit
mixture. Serve over ice; top with cava, if desired.* SERVES 6

THE SPIRITS *of the* CACTUS FRUIT

COLONCHE	A fermented drink made from the juice or pulp of the prickly pear cactus, *Opuntia* spp.
NAVAI'T	A fermented winelike drink made from the fruit of the majestic saguaro, *Carnegiea gigantean.*
PITAHAYA OR PITAYA	A wine made from the fruit of either the organ pipe cactus *Stenocereus thurberi* or various species of *Hylocereus*, also known as dragon fruit.

BUGS *in* BOOZE: COCHINEAL

--- *Dactylopius coccus* ---

The prickly pear cactus made another important contribution to the world of spirits and liqueurs: carmine dye. The white fuzzy pest found on *Opuntia* cacti is actually a scale called cochineal. Scale are sucking insects that latch onto a plant and feed off its sap, hiding under a waxy cover that makes them resemble a tick. The cochineal scale are particularly easy to spot because they cover themselves in a fuzzy white material to hide their offspring and protect them from drying out. Underneath that white fluff, the insects secrete carminic acid, a defensive chemical that deters ants and other predators and also happens to be bright red.

In Mexico, Spanish explorers wondered about the vivid red dyes that native people used on blankets and other textiles. At first they thought the color came from the red prickly pear fruit itself. Fernández de Oviedo, writing in 1526, claimed that eating the fruit turned his urine bright red (which was either a complete falsehood or a sign of a much more serious medical problem). They soon learned that the dye came from cochineal. To make it, the scale would be scraped off the cactus, dried, and then mixed with water and alum, a kind of natural fixative. The Spanish had some experience with the use of bugs as dye—they'd been using another scale, *Kermes*, for a similar purpose—but cochineal produced a much more vivid red.

Since the 1500s, cochineal-based carmine dye has been used as a coloring for confections, cosmetics, textiles, and liqueurs. It gave Campari its rich red color until 2006, when company officials say it was removed owing to supply problems. Reports of people with allergies going into anaphylactic shock, as well as a general squeamishness about insect ingredients in food, have led to new labeling requirements in the United States and the European Union. In the EU, any product colored with cochineal must state it on the label. It may be called E120, Natural Red 4, or carmine or cochineal. (The color once came from Polish scale, *Porphyrophora polonica*, as well, but it is severely endangered and no longer used.) In the United States, labels must read "cochineal extract" or "carmine."

SAVANNA BAMBOO

Oxytenanthera abyssinica (syn. *O. braunii*)

POACEAE (GRASS FAMILY)

Also called wine bamboo, this fast-growing member of the grass family is used for fencing, tools, basketry, erosion control—and alcohol. In Tanzania, the young shoots are cut and then bashed twice a day for a week, injuring the plant to encourage the sap to flow. It ferments naturally in as little as five hours. The bamboo wine, called *ulanzi,* is only made during the rainy spring season when the young bamboo is growing. Women make batches of it to sell by the liter in their villages. It is not uncommon for travelers to get a free sample as they walk from one village to the next: the stands of bamboo are left unattended while the sap flows into containers. The temptation to simply help oneself to a drink along the journey is hard to resist.

STRAWBERRY TREE

Arbutus unedo

ERICACEAE (HEATH FAMILY)

The red, rough-skinned fruit of the strawberry tree, perfectly round and about the size of a cherry, is not nearly as tasty as the fruit it is named after. In fact, botanists say that the species name *unedo* comes from the Latin *unum edo,* meaning "I eat one." Just one.

But distillers—most of them unlicensed and working on equipment that could have come straight from the Middle Ages—turn the fruit into a popular local spirit called *aguardiente de medronho.* Although it is available commercially, it is more commonly shared among families and sold to neighbors, particularly in the Algarve region of southern Portugal.

Rather than bloom in spring, as most fruit-producing trees would, the strawberry tree blooms in fall, at the same time that the previous year's fruit is ripening. In Portugal and Spain, that process begins in September. Pickers gather only the ripest fruit, returning once a month through December to complete the harvest.

Once picked, the fruit is mashed or submerged whole in water and fermented for three months. Then, usually in February, it is boiled over a wood fire and distilled in a copper alembic still, with a pipe running through a barrel of water that serves as the condenser. The result is a high-proof spirit, usually above 45 percent ABV, which is either bottled immediately or aged in oak for six months to a year. In Spain, a sweeter, lower-proof liqueur called *licor de madroño* is made by macerating the fruit in high-proof spirits with sugar and water.

The strawberry tree is a type of madrone, one of fourteen species found throughout Europe and North America. Most madrones are small, beautiful trees with glossy, narrow leaves and a reddish, peeling bark. None of them produce particularly tasty fruit, in spite of the fact that they are relatives of the blueberry, huckleberry, and cranberry. *A. unedo* is nonetheless grown in warm climates around the world as an ornamental. A cultivar called Elfin King is even grown in containers and is considered to produce tastier fruit than most.

THIS BEAR IS NOT DRUNK

Madrid's coat of arms shows a bear standing on its hind legs, eating fruit from the strawberry tree. A statue depicting this scene can also be found in the city's center, at the western end of the Puerta del Sol. While locals like to claim that the bear is getting drunk from the fermenting fruit of the tree, the fruit does not, in fact, ferment on the tree to such an extent that it could intoxicate an animal as large as a bear. This appears to be yet another tall tale of animal intoxication.

TAMARIND

Tamarindus indica
FABACEAE (BEAN FAMILY)

The tamarind probably originated in Ethiopia and found its way to Asia via ancient trade routes. Today it grows in tropical climates all over the world, most notably in East Africa, Southeast Asia, Australia, the Philippines, in Florida, and throughout the Caribbean and Latin America.

The tree reaches up to sixty feet in height, with a canopy of small, feathery leaves that throw off much-needed shade. The fruit is actually a long seedpod with a slightly sweet, slightly tart, edible brown pulp. It is used in curries, pickles, candies, and as a flavoring in Worcestershire sauce, where it might then make an appearance in Bloody Marys or Micheladas, a Mexican drink that combines beer with tomato juice (or Clamato), lime juice, spices, and sauces. Although there are over fifty different cultivars, they are difficult for anyone but a local to tell apart. Tropical plant nurseries mark them simply as "sweet" or "sour" varieties. The sweet variety is eaten raw, but it is actually the sour variety that is used in drinks and for cooking.

Tamarind wine is made removing the dry, outer husk of the seed pod, scooping out the pulp and pressing the juice from it, and then fermenting a mixture of juice, water, and sugar. The wine can be found today in the Philippines, particularly in Batangas, just south of Manila. Tamarind also turns up as a flavoring in liqueurs, like Mauricia Tamarind Liqueur, a rum-based drink from the island of Mauritius south of Madagascar in the Indian Ocean. Tequila distillers have created *licores de tamarindo* as well. Tamarind paste or syrup, available at specialty food markets, is becoming a popular cocktail mixer, particularly in margaritas, where it hits the same sweet-sour notes that lime juice does.

PART II

We Then Suffuse Our Creations
with a Wondrous Assortment of Nature's Bounty

There's more in the average liquor bottle than straight alcohol. Once a spirit leaves the still, it is subject to endless experimentation with herbs, spices, fruits, nuts, bark, roots, and flowers. Some distillers claim to use over a hundred different botanicals in their secret recipes. Here are just a few of the plants you're likely to find in tonight's cocktail.

-- we begin with --

HERBS & SPICES

herb:

THE TENDER, GREEN VEGETATIVE
OR FLOWERING PART OF A PLANT
USED FOR FLAVORING.

spice:

THE DRY, TOUGH WOODY PARTS
OF A PLANT (SUCH AS BARK, SEEDS,
STEMS, ROOTS) AND, IN SOME CASES,
FRUIT, USED FOR SEASONING.

ALLSPICE

Pimenta dioica

MYRTACEAE (MYRTLE FAMILY)

Classic cocktail aficionados are accustomed to finding strange and unfamiliar ingredients in old recipe books, but few are more confusing than pimento dram. A drink made of those rubbery red things stuffed inside olives? What could that possibly taste like?

Fortunately, pimento dram is not made of the pimento found in an olive. It is a liqueur made from rum, sugar, and allspice. And the reason allspice and mild red peppers share a name is an accident of history.

Spanish explorers traveling to the West Indies and Central America saw people adding small dark berries to their traditional meals and to chocolate. They seemed to add heat and spice to the dish, so the Spaniards assumed they were some kind of pepper. For that reason, they called the plant pimento, their word for pepper. In 1686, British naturalist John Ray described it in his monumental three-volume work *Historia Plantarum* as "sweet scented Jamaica pepper." And, because it could be used in such a wide variety of dishes, he also called it all-spice.

The allspice tree flourishes in tropical regions of the Americas and in Jamaica. It produces pea-shaped berries that each hold two seeds. The berries are picked green in midsummer and spread on the ground to dry in the sun, or gently heated in ovens. The flavor is similar to that of cloves, and in fact the two trees, which are closely related, both produce the aromatic oil eugenol.

Early spice traders tried to plant allspice seeds around the world but found them impossible to germinate. Eventually it was discovered that the seeds must pass through the body of a fruit-eating bat, a baldpate pigeon, or some other local bird, in order to be sufficiently heated and softened for germination. Today, through the agency of birds, the tree has become invasive in Hawaii, Samoa, and Tonga.

The near devastation of the world's allspice trees occurred in the Victorian era, when trees were cut down not for their spices but for their wood. It was all the rage to manufacture umbrellas and walking sticks from the pale, aromatic sticks because they resisted bending or cracking. Millions of trees were destroyed. To protect

THE BAY RUM

An extract from the leaves and berries of *Pimenta racemosa*, a close relative of the allspice tree, is added to high-proof Jamaican rum to make bay rum cologne. Although the ingredients sound delicious (and people who wear it smell delicious), the concentrated botanical extract delivers an unusually high dose of eugenol that would be toxic if imbibed. Wear the cologne, but drink a *Pimenta* tree in this form instead. This drink is sweet but not childish, and gives off the pale, tangerine glow of a Caribbean sunset. Velvet Falernum is a wonderful spicy, syrupy mixer from Barbados available at better liquor stores, but if you don't have any, simple syrup will do.

1½ ounces dark rum
½ ounce St. Elizabeth Allspice Dram, or another pimento dram
½ ounce Velvet Falernum or simple syrup
Dash of Angostura bitters
Fresh juice of 1 orange or tangerine segment (feel free to experiment with lime or other citrus as well)

Shake all the ingredients over ice and serve on the rocks in an Old-Fashioned glass.

them, Jamaica enacted a strict ban on the export of allspice saplings in 1882.

Allspice is an ingredient in perfumes and liqueurs. It is sometimes found in gins and is believed to be part of the mysterious formulas of Benedictine and Chartreuse, as well as other French and Italian cordials.

Pimento dram, also called allspice dram, is an ingredient in classic tiki cocktails and has recently become popular in warm, spiced autumnal drinks, where it gives a baked spice flavor to Calvados or apple brandy.

✤ ALOE ✤

Aloe vera
ASPHODELACEAE (ALOE FAMILY)

Like its cousin, the agave, the aloe is sometimes mistaken for a cactus. In fact, it is more closely related to lilies and asparagus than cactus. But like a cactus, it does love heat and dry weather. And while people who drink aloe juice might never guess it, aloe contains one of the most bitter flavors in the world. For this reason it turns up in more than a few bottles behind the bar.

Aloe traveled from its native sub-Saharan Africa to Asia and Europe in the seventeenth century. Today nearly five hundred species have been identified, and they span the globe, growing in tropical climates where winter temperatures stay above 50 degrees Fahrenheit.

Like other succulents, aloes depend upon a special kind of photo-synthesis that requires them to open their pores—stomata—only at night to breathe. They take in carbon dioxide and store some to use the next day, which allows them to essentially hold their breath all day long. When they do breathe, they release as little water as possible through those pores, relying on cooler nighttime temperatures to slow the loss of water.

And of course, they store water in their leaves, which explains the thick, juicy gel familiar to anyone who has ever attempted a little first aid in the outdoors. While the gel is useful to protect wounds—a latex made by the plant covers the wound while allowing it to breathe—its use as an internal medicine is not entirely proven. Some species are even poisonous, which makes it important to think twice before ingesting an unfamiliar aloe.

The bitter component in aloe, called aloin, is found in the latex just under the surface of the leaf. Scientists have recently learned that a particular gene allele makes some people highly sensitive to aloin's bitterness, whereas people without that allele can't even taste it except at high concentrations. This may explain why some people love Italian bitters, also called *amaro*s, and others can't stand them.

Aloe is one of the ingredients that gives Fernet-style *amaro*s, such as Fernet Branca, their bracing quality. While quinine, gentian, and a number of other plants are also used to impart bitterness, they also give a slightly vegetal or even floral flavor. Aloe brings with it no such extra notes. If bitterness had a color, aloe would be black as coal.

To make juice from aloe, the liquid is extracted from the center of the leaf and filtered to remove the aloin and the dark color it imparts. This makes it more palatable and perhaps safer: Aloin was once an ingredient in laxatives, but during a routine evaluation of ingredients that had never undergone modern safety reviews, the U.S. Food and Drug Administration (FDA) banned its use in laxative products—not because it had been proven dangerous but because no pharmaceutical company offered to demonstrate its safety and effectiveness using modern methods. Still, its traditional use as a laxative may explain why aloe's bitter components were used in the formulas for digestifs.

ANGELICA

Angelica archangelica
APIACEAE (CARROT FAMILY)

Angelica, a medieval herb native to Europe, is a flavor that seems to turn up in the secret formulations of Chartreuse, Strega, Galliano, Fernet, vermouth, and perhaps even Benedictine and Drambuie. The dried root was an early remedy for digestive problems.

Angelica is related to parsley and dill, which is where it gets its bright, refreshing, and decidedly green flavor. It is also related to poison hemlock and a number of other toxic plants. In fact, of the over twenty-five species of angelica, many have not been evaluated for toxicity, and some closely resemble their more poisonous relatives, making it a risky plant to collect in the wild. Fortunately, the edible *Angelica archangelica*, sometimes sold as *A. officinalis,* is widely available from nurseries and seed companies. It is usually grown from seed because plants with a long taproot, like angelica, don't transplant well. The plant reaches six feet tall and makes a striking statement with its large, finely toothed leaves and white umbel-shaped flowers similar to Queen Anne's lace.

While the stems have been used to make candied angelica, it is the seeds and dried roots that flavor wines and liqueurs. Angelica is a biennial, which means that seeds take two years to germinate, grow, and become mature plants that produce flowers and another generation of seed. If it is being grown for its roots, it would generally be harvested in the autumn of its first year, while the root is still tender and hasn't been colonized by bugs. (Some are allowed to overwinter and bloom the second year for seed stock.) A chemical analysis of fresh angelica root shows that it contains a number of tasty compounds designed to ward off insect attacks: citrusy limonene, woodsy pinene, and the distinctly herbal β-phellandrene are all flavors that make it particularly welcome in liqueurs.

THE JOYS OF LIQUORE STREGA

WHILE THE YELLOW ITALIAN LIQUEUR STREGA can be mixed into a cocktail—it plays well with gin in martini variations, for instance— there is no reason to bother with that. Strega is divine on its own.

Its manufacturers claim that the recipe dates to 1860, when it was given the name Strega, meaning "witch," to refer to the legendary witches of the town of Benevento, just south of Naples. The distillery is still there today.

Strega is a sweet, complex herbal liqueur that is perfect after dinner, served neat or on the rocks. The distiller admits to a few of its seventy ingredients: cinnamon, iris, juniper, mint, citrus peel, cloves, star anise, and myrrh, along with saffron for color. Visitors to the distillery have also reported seeing cloves, nutmeg, mace, eucalyptus, and fennel. But angelica is widely believed to be one of its primary flavors. Taste it and decide for yourself.

ARTICHOKE

Cynara scolymus (syn. *Cynara cardunculus* var. *scolymus*)
ASTERACEAE (ASTER FAMILY)

The artichoke got its start as a cardoon. This leafy ancestor, *C. cardunculus,* probably originated in north Africa or the Mediterranean. It was actively cultivated by Egyptians, Greeks, and Romans, and through their efforts a separate species, the artichoke, emerged. The two plants look very similar, with long, silvery, deeply serrated leaves and thistlelike flowers. The two can even interbreed if planted closely together. Cardoon stalks were considered both food and medicine, while artichokes were cultivated more for their oversized flower buds. Both plants spread throughout Europe in the fifteenth century and became an important part of Italian cuisine.

Artichokes and cardoons have a long history as ingredients in digestive tonics. In fact, recent research shows they may stimulate bile production, protect the liver, and lower cholesterol levels. The active compounds are cynaropicrin and cynarin, both of which are found in higher levels in the leaves. Artichokes also play a well-known trick on the taste buds, temporarily suppressing taste receptors on the tongue that detect sweetness. The next thing to come across the palate—a drink of water, a bite of food—tastes unusually sweet as those receptors start working again. This makes artichokes notoriously difficult to pair with wine, but the strange blend of bitter and sweet is perfect in a cocktail.

Several Italian *amaro*s rely on artichoke and cardoon. The aptly named Cynar is the best example; it's wonderful on its own or in soda, and works very well as a Campari substitute in a Negroni. Cardamaro Vino Amaro, made in Italy's Piemonte region, is a wine-based infusion of cardoon, blessed thistle, and other spices. It has a lower alcohol content (17 percent ABV) and an oxidized sweetness similar to that of sherry or sweet vermouth. Other regional versions are usually labeled simply "Amaro del Carciofo."

BLESSED THISTLE: BECAUSE ONE GREAT THISTLE DESERVES ANOTHER

The word *thistle* is not a botanical term; it's more of a popular word used to describe plants with prickly leaves and spiky flowers atop a round, bulbous base. Artichokes and cardoons are often called thistles, but a close relative actually goes by the name blessed thistle, or *Centaurea benedicta*. The two foot-tall, yellow-flowered herb resembles a hairy dandelion—and like dandelions, it is both weedy and bitter. All parts of the plants are used in digestive tonics, vermouths, and herbal liqueurs; the active ingredient seems to be a compound called cnicin, which is being evaluated for its anti-tumor properties.

BAY LAUREL

Laurus nobilis

LAURACEAE (LAUREL OR AVOCADO FAMILY)

The leaves of this Mediterranean tree were once used to fashion a crown for winners of Greek and Roman sporting events, but they are also used to flavor stews, sauces, and meat dishes. The small black berries are an ingredient in traditional French cooking. The tree's essential oils include eucalyptol, which explains its strong eucalyptus essence. Smaller quantities of linalool and terpineol give it a green, spicy, pungent, and piney taste.

Bay laurel infuses vermouths, herbal liqueurs, *amaro*s, and gins. A French distiller, Gabriel Broudier, makes a pear and bay leaf liqueur called Bernard Loiseau Liqueur de Poires Laurier. The Dutch liqueur Beerenburg contains a distillate of laurel leaves with gentian and juniper berries.

The California bay laurel (also called the Oregon myrtle), *Umbellularia californica,* is sometimes used as a substitute. However, other plants that are called laurels, including the cherry laurel (*Prunus laurocerasus*) and the mountain laurel (*Kalmia latifolia*) are extremely poisonous—so a home infusion of just any random laurel plant would be ill advised. Fortunately, the real bay laurel grows throughout Europe and in parts of North America, and both leaves and berries are sold as kitchen spices.

BETEL LEAF

Piper betle
PIPERACEAE (PEPPER FAMILY)

This small, dark green vine, a close relative to the vine that produces black pepper, is best known as the wrapper in which the betel nut, *Areca catechu,* is placed. The two comprise a little bundle known as a quid or *paan.* The combination delivers a mild, addictive stimulant enjoyed by four hundred million people around the world, primarily in India and Southeast Asia. Unfortunately, the quid also causes cancer, turns teeth black, and produces a steady flow of red saliva that is often spat out on the street.

The betel leaf is also used to wrap other things. A "sweet *paan*" refers to a betel leaf filled with fruit and spices; it might be served to guests after dinner as a (nonstimulating) desert. Betel leaves can also be filled with tobacco, another custom that alarms public health officials for the high rates of oral cancer it causes.

Paan liqueur is made in Sikkim, a region that borders Nepal. While home brewers and commercial distillers are equally reluctant to divulge their recipes, locals are certainly under the impression that they are drinking a spirit steeped in or distilled with the betel leaf, and perhaps the nut as well. A few *paan* liqueurs are distributed internationally, and while the distillers make no claim about ingredients, it is unlikely that that the betel leaf is used in those versions. Neither the leaf nor the nut are approved food ingredients in the European Union or in the United States. In fact, both are included in the FDA's database of poisonous plants. (This is not to say that they are illegal to grow; a few tropical nurseries sell them.) In 1995, the *Los Angeles Times* reported on the launch of Sikkim Paan Liquor, which apparently contained no betel leaf at all, but cardamom, saffron, and sandalwood, bringing to mind a combination of Drambuie and an Indian spice shop.

The leaf may prove to have some redeeming qualities. A 2011 medical study published in the journal *Food & Function* investigated several spices for possible protective effects against alcohol-induced liver damage. A number of Indian spices and herbs looked promising, including turmeric, curry, fenugreek, tea—and the leaves of *Piper betle*.

BISON GRASS

Hierochloe odorata
POACEAE (GRASS FAMILY)

This tough, perennial grass, also called sweetgrass, is prized for its vanilla-like fragrance. It is native to both North America and Europe and has been used by Native Americans to make baskets and incense. In Poland, it is an ingredient in a traditional flavored vodka called *żubrówka.* A wild stand of the grass still grows in the Białowieża Forest, between Poland and Belarus, where it feeds a herd of wisent, the endangered European bison.

A limited amount of the wild grass can be gathered every year to make *żubrówka.* Once harvested, it is dried and macerated in rye vodka. A single blade of grass floats in every bottle. The spirit has been unavailable in the United States since 1954 because the grass contains coumarin, a banned substance that can be turned into a blood thinner in the laboratory or in the presence of certain species of fungus. While the conversion of coumarin to a blood thinner is easily avoided, the ban on anything containing coumarin remains. Recently Polmos Białystok, the makers of *żubrówka,* found a way to remove the coumarin, making it legal once again in the United States.

The traditional way to drink it is to mix one part *żubrówka* to two parts clear, cold apple juice. This recipe is simply a variation on that tradition:

BISON GRASS COCKTAIL

1½ ounces *żubrówka*
½ ounce **dry vermouth**
½ ounce **apple juice**

Shake all the ingredients over ice and strain into a cocktail glass.

CALAMUS (SWEET FLAG)

Acorus calamus
ACORACEAE (SWEET FLAG FAMILY)

S weet flag is a highly fragrant grass or rushlike plant that grows in marshy areas throughout Europe and North America. The rhizome has a complex, spicy, bitter flavor that lends itself to *amaros* like Campari and herbal liqueurs like Chartreuse, as well as gin and vermouth. The flavor has been described as woodsy, leathery, and also creamy; perfumer Steffen Arctander described it as smelling like a milk truck or the inside of a shoe repair shop.

Some varieties of the plant contain a potentially carcinogenic compound called β-asarone. For this reason, the FDA has banned it as a food additive. However, not all sweet flag is equally dangerous. The American variety, called *A. calamus* var. *americanus* or *A. americanus,* does not have any significant quantity of the potential toxin, and European strains also have relatively low levels. The European Union acknowledges that the plant is widely used in bitters, vermouths, and liqueurs, and has set limits regulating the amount of β-asarone in alcoholic beverages and encouraging the use of less toxic varieties. In the United States, distillers sidestep the ban by producing liqueurs that contain undetectable levels of the toxin.

CARAWAY

Carum carvi

APIACEAE (CARROT FAMILY)

Norwegian distillers don't recount myths of lost princes and ancient recipes to explain the mysterious origins of their classic spirit. Instead, they tell the story of a trading expedition gone wrong. According to the makers of Linie Aquavit, a trading ship bound for Indonesia in 1805 carried used sherry casks filled with the caraway-flavored spirit aquavit in its cargo hold. The traders were unable to sell their national drink in Indonesia and returned home with it.

When they arrived in Norway, they found that the long, rollicking sea voyage had greatly improved the aquavit. To reproduce the flavor, they tried simply storing the aquavit in sherry casks, but it wasn't the same. The brutal sea voyage, with its time in warm equatorial seas and in cold Nordic waters, combined with the tossing and turning of the ship, caused the casks to expand and contract in a way that released more flavor from the oak. For that reason, casks of Linie still voyage around the world for four and a half months on the decks of cargo ships, crossing the equator twice and visiting thirty-five countries. The distiller once kept this strange method of aging the spirit a secret, but now a log of the voyage is printed on every label.

Aquavit is flavored with caraway, an annual herb that is a close relative of parsley and cilantro. What people refer to as the seed is actually a fruit that contains two seeds, along with essential oils that give it its spicy, toasted flavor. Most people associate the flavor with rye bread, but it is also used in sauerkraut, coleslaw, and some Dutch cheeses.

Caraway is native to Europe. Archeological evidence in Switzerland points to the use of the seeds as a spice as many as five thousand years ago. There are two types: a biennial winter type that is sown in spring or fall for a harvest the following winter and an annual type that is sown in spring for a fall harvest. The winter type is the traditional choice in eastern Europe and is most widely available from seed companies.

Aquavit is made with a potato vodka base. Caraway is the predominant flavor, but fennel, dill, anise, cardamom, cloves, and citrus might also be added. Another caraway-based spirit is allasch, a Latvian liqueur also made with anise, and the better-known kümmel, a sweet, grain-based liqueur that dates to sixteenth-century Holland, which is usually served on the rocks after dinner.

CARAWAY/CUMIN CONFUSION

Caraway and its close relative cumin (*Cuminum cyminum*) are often confused with one another, even though cumin has a much stronger, more peppery flavor. The common name for the two plants has historically been the same or nearly identical in many eastern European languages. In Germany, for instance, cumin is *Kreuzkümmel* and caraway is *Kümmel*. Although cumin is one of the world's most popular spices, it is not widely used to flavor spirits.

CARDAMOM

Elettaria cardamomum var. *Minor* or var. *Major*
ZINGIBERACEAE (GINGER FAMILY)

If you've never seen a cardamom plant, picture a clump of tall, weedy orchids. As a member of the ginger family, cardamom produces the third most expensive spice in the world, after saffron and vanilla. Its high price comes in part from the tropical locations it prefers and in part because the fruit is painstakingly difficult to harvest.

Cardamom has been collected in the wild for hundreds of years but was brought into cultivation in the nineteenth century. The plant reaches nearly twenty feet in height and blooms over a long season, requiring pickers to return again and again to the same plant to harvest individual fruits. They must be picked while they are still slightly green, then dried and split carefully apart to remove the seeds within. The pods are also sold intact, with the seeds still inside, which preserves more of the flavor.

Cardamom from India is considered the best quality, although Guatemala has become a major producer as well. There are two types: the Malabar type has a slight eucalyptus flavor, while the Mysore type is warmer and spicier, with citrus and floral notes. A related species, *Amomum subulatum,* also called large cardamom or black cardamom, is typically dried on an open fire and has a much smokier flavor as a result.

The spice contains high levels of linalool and linalyl acetate, which are fragrant compounds also found in lavender, citrus, and a wide range of other flowers and spices. Japanese scientists recently showed that these compounds reduce stress, as measured by direct testing of subjects' immune system response. That's as good a reason as any to add it to a drink.

Cardamom flavors a wide range of spirits, including gins, coffee and nut liqueurs, vermouth, and Italian *amaro*s. The best way to use it in a cocktail is to heat green cardamom seeds with simple syrup and experiment with it in a wide range of spicy, tropical, and fruit-based drinks.

CLOVE

Syzygium aromaticum
MYRTACEAE (MYRTLE FAMILY)

A clove is not a seed or a fruit or even the bark of a tree. It is, in fact, a tightly closed flower bud that has been plucked from an Indonesian tree and spread out in the sun to dry before it ferments (in the way that, seemingly, anything will ferment if left unattended.)

Cloves come from the Indonesian spice islands of Ternate, Tidore, Bacan, Makin, and the Maluku Islands, which have been the source of spices for Asia and Europe since at least the third century BC. The Romans eagerly traded with Arab merchants for exotic botanicals from these islands, and by the seventeenth century, the Dutch and Portuguese were fighting over the territory. In an attempt to control the market, the Dutch cut down clove trees on all but the islands they controlled. French and British traders eventually got hold of some clove seedlings and exported them to their own tropical colonies, including Sri Lanka, India, and Malaysia. Sadly, this had the effect of wiping out the rich genetic diversity that may have once existed among wild clove trees. The only wild trees that remain contain no trace of eugenol, the distinctive flavor extracted from modern cloves. This suggests that a second wild ancestor, which did produce eugenol, was wiped out entirely by the spice traders.

The clove tree itself is quite beautiful; the leaves transition from pale gold to pink to green throughout the season. The buds also change color as they bloom and must be harvested at the precise moment they turn light pink. Because of the tree's long blooming cycle, the flowers are picked as many as eight times in a season, yielding only about ten pounds of cloves per year. Clove stems are sometimes used as a cheap substitute for the buds, and clove oil may also be extracted from the leaves and branches.

The varieties of cloves sold in the trade today are Zanzibar, Siputih, and Sikotok, with the Siputih being the largest and most pungent of the three. Clove extract has been used throughout history—and continues to be used today—as a dental anesthetic because of its numbing and analgesic effects. In fact, that distinctive dentist office smell comes in part from cloves.

However, there are much more pleasant ways to enjoy cloves than a trip to the dentist's office. The flavor is wonderful in combination with other spices. It intensifies vanilla flavors and adds a level of complexity to citrus. Many nutty and spicy liqueurs rely on cloves to support and amplify other flavors, including amaretto, alkermes, and some vermouths and *amaro*s.

COCA

Erythroxylum coca

ERYTHROXYLACEAE (COCA FAMILY)

No plant is more symbolic of our endless war on drugs than this small, dark green Andean bush. When the leaves are chewed, they act as a gentle stimulant and may offer protection against altitude sickness. Archeologists have found evidence that Peruvians used the plant in this manner as early as 3000 BC, and they were still using it when the Spaniards showed up in the sixteenth century. The Catholic church tried to ban it but quickly realized that enslaved Peruvians could be made to work harder if they had their coca, so it remained a part of the culture.

Europeans, always on the search for a new plant that could be put to some medical or recreational use, found a way to extract the pure cocaine alkaloid, creating a drug with a much more powerful effect than the leaves alone. Cocaine became a pain reliever, antiseptic, digestive tonic, and all-around cure. Freud liked it; in 1895 he wrote that "a cocainization of the left nostril had helped me to an amazing extent."

The leaves were used in wines and tonics as well, the most famous being the French Vin Mariani, whose advertisements promised that it was an "Effective and Lasting Renovator of the Vital Forces." In 1893, the company published a charming illustrated book of testimonials for its product that began with an introduction about the coca plant ("Not Cocoa or Cacao," it emphasized), in which it claimed that "the most effective form of administering Coca is the vinous one."

The commendations came from celebrities like Sarah Bernhardt, who declared that the wine "helped to give me that strength so necessary in the performance of the arduous duties which I have imposed upon myself." The French cardinal Charles Lavigerie, who oversaw missionaries in Africa, wrote that "Your coca from

America gives to my 'White Fathers,' sons of Europe, the courage and strength to civilize Asia and Africa." The best endorsement came from controversial French politician Henri Rochefort, who said that "Your precious Vin Mariani has completely reformed my constitution; you ought certainly offer some to the French Government."

Coca continues to flourish in its native range in the Andean mountains. The shrubs grow to about eight feet in height, producing small white flowers and seeds. Only the young, fresh leaves are harvested, usually three times a year, beginning with the rainy season in March. There are seven species in all, including at least one other, *Erythroxylum novogranatense,* that also contains the cocaine alkaloid. *E. rufum,* or false cocaine, is entirely free of the alkaloid and is grown by some botanical gardens in the United States.

Although wine, tonic, and soda manufacturers are no longer allowed to include cocaine in their formulations, they can still use a cocaine-free flavor extract from the plant. The FDA has approved "Coca (decocainized)" as a food additive, and one American manufacturer, the Stepan Company in New Jersey, has been granted the license to legally purchase the leaves from Peru's National Coca Company. It separates the cocaine alkaloid for use as a topical anesthetic, and sells the remaining flavoring to companies like Coca-Cola. Not to be outdone, the Bolivian government has funded the creation of a number of coca-flavored sodas and other products, arguing that it is hypocritical for the United States to sanction the use of the leaves in American soft drinks while frowning on local products made from the same plant.

Although it is entirely legal to flavor liquor with the decocainized extract of coca leaves, few distillers do so. One notable example is the herbal liqueur Agwa, which is widely sold in the United States and Europe with a dramatic label announcing its controversial ingredient. (Other ingredients include guarana seeds, a South American vine with a caffeine-like compound, and ginseng.) In coca-producing countries, *licor de coca* and *vin de coca* are also sold in local markets.

CORIANDER

Coriandrum sativum

APIACEAE (CARROT FAMILY)

Coriander is a favorite ingredient among distillers. It's found in almost all gins and in many herbal liqueurs, absinthe, aquavit, pastis, and vermouth. But anyone who has ever eaten the leaf of a coriander plant—called cilantro in the Americas—might wonder why they so rarely encounter that distinctive flavor in any of these drinks.

The reason is that the fruit—round, brown seeds—undergo a chemical change as they dry, shedding that bright cilantro flavor completely. The essential oil found in the fresh leaves and on the surface of the unripe fruit is instantly recognizable and not to everyone's liking, owing to genetic differences in how people perceive flavors. Some people call it fetid; others say it smells of bugs. In fact, the Greek word for bed bug, *koris,* is the root of its early Greek name *koriandron.*

But deep inside the fruit is another oil that is easily extracted once the fruit is dry and the characteristic cilantro flavor has evaporated. That oil, which is dominated by linalool, thymol, and geranyl acetate, a compound found in geraniums, is the perfect blend for booze. It combines the woodsy notes of thyme, the rich perfume of geranium, and the bright, floral, citrus flavor of linalool. It tastes, in other words, like very good gin.

Two varieties turn up in the spice market: the high-quality Russian coriander *C. sativum* var. *microcarpum,* which is smaller but higher in essential oil, and the larger-fruited *C. sativum* var. *vulgare,* sometimes referred to as Indian, Moroccan, or Asiatic coriander. The latter is grown for its leaves and is more widely available to gardeners. (Many varieties sold to gardeners have been bred to resist bolting, or setting seed, so they will produce more leaves for cooking.) The best-quality oil séems to come from plants grown in areas with cool, wet summers, which is why both Norway and Siberia supply top-notch coriander to the world market.

CUBEB

Piper cubeba

PIPERACEAE (PEPPER FAMILY)

This climbing, woody Indonesian vine produces a fruit that was once more popular than its better-known relative, *Piper nigrum,* or black pepper. Although the dried fruits look like black pepper, they are usually sold with their stems attached, making the two easy to tell apart. Its pungent bite comes from a compound called piperine, although it actually contains even higher levels of limonene, that ubiquitous flavor found in a wide range of citrus and herbs. This may help explain why cubeb is so popular as an ingredient in gin, where spice and citrus enjoy a happy marriage.

In the Victorian era, "medicated" cubeb cigarettes were sold as asthma treatments. Modern cigarette companies who have published their ingredients still list it as a flavor. Seventeenth-century Italian priest Ludovico Maria Sinistrari, who wrote extensively on the use of plants to perform exorcism, prescribed a brandy-based tonic flavored with cubeb, cardamom, nutmeg, birthworts, aloe, and other roots and spices to keep the demons away.

☙ DAMIANA ☙

Turnera diffusa
TURNERACEAE (DAMIANA FAMILY)

In 1908, federal officials confiscated a bottle labeled "Damiana Gin" that was being shipped from New York to Baltimore. The label advertised the aphrodisiac qualities of the spirit, but the feds had their suspicions. Laboratory analysis turned up strychnine and brucine (both poisons derived from the strychnine tree) as well as salicylic acid, an aspirin-like compound extracted from willow trees that can be dangerous in large doses.

Given its toxic ingredients, the "false and misleading" claims of the drink's aphrodisiac qualities, and the fact that it was not actually gin, the bottle was deemed to be in violation of the 1906 Pure Food and Drug Act. Its owner, a man named Henry F. Kaufman, was fined one hundred dollars for shipping a product that violated the act. But damiana's reputation persists.

This six-foot-tall, highly aromatic shrub produces tiny yellow flowers and small fruits. It grows wild in Mexico, where it has a reputation for stimulating the sexual appetite. In the nineteenth century, physicians prescribed it as a sexual tonic; one doctor writing in 1879 said that it could be given to female patients "to produce in her the very important yet not absolutely essential orgasm."

Remarkably, there may be something to these claims. A 2009 study showed that the plant could speed recovery time in "sexually exhausted male rats," allowing them to perform a second feat of lovemaking after a very short interval. (The method used to sexually exhaust the rats was not disclosed.)

In spite of this intriguing bit of research, no clinical trials have been conducted to determine the plant's effects on humans. It is a legal food additive in the United States, and the dried leaves and stems flavor Damiana, a Mexican herbal liqueur sold in a bottle shaped like—what else?—a fertility goddess.

DITTANY OF CRETE

Origanum dictamnus
LAMIACEAE (MINT FAMILY)

The mysterious-sounding dittany of Crete is nothing more than an odd-looking type of oregano. The round, silver, fuzzy leaves and bracts of pinkish purple flowers make it a showstopper in Mediterranean gardens, which is why its habitat is no longer limited to a single Greek island. It has earned the name hop marjoram because the flowers resemble hops, but the plant's fragrance is more similar to thyme and other oreganos. The leaves have been used to flavor medicinal tonics since at least early Greek times, and today they are still used in vermouths, bitters, and herbal liqueurs.

ELECAMPANE

Inula helenium
ASTERACEAE (ASTER FAMILY)

A wild stand of elecampane could easily be mistaken for a bunch of overgrown dandelions—and in fact, the two plants are related. Although elecampane is native to southern Europe and parts of Asia, it now grows wild throughout much of North America, Europe, and Asia and is cultivated and sold as a medicinal herb to treat coughs. It reaches eight feet tall and sports small, daisy-shaped yellow flowers. The bitter, camphor-flavored root is a common ingredient in vermouths, bitters, absinthe, and herbal liqueurs.

EUROPEAN CENTAURY

Centaurium erythraea

GENTIANACEAE (GENTIAN FAMILY)

This pink-flowered annual herb is a relative of gentian. Native to Europe, it has spread to North America, Africa, and parts of Asia and Australia. The dried stems and leaves have historically been used externally to treat wounds and internally as a digestive tonic. Today the plant's bitter iridoid glycosides—powerful compounds that the plant uses to defend itself—make it useful as an ingredient in bitters and vermouths.

FENUGREEK

Trigonella foenum-graecum

FABACEAE (BEAN FAMILY)

Starting in 2005, people living in certain parts of New York City would suddenly develop the strangest craving for pancakes. A distinctly maple syrup smell was wafting through town. It didn't happen very often, but when it did, people called the city to inquire as to the source of the unexplained, but not altogether unpleasant, odor.

Finally, in 2009, city officials had an answer: fenugreek. The seeds of this diminutive beanlike plant are ground and mixed into curry spices—but they are also processed by a company in New Jersey that sells industrial fragrances and flavors. The caramel or maple syrup note contributed by fenugreek is used as a flavoring in liqueurs as well as in imitation maple syrup and other sweets.

Fenugreek comes from the Mediterranean, northern Africa, and parts of Asia; it has been a traditional part of Indian and Middle Eastern cooking for centuries. While it never plays the starring role in a liqueur, it may be used in the background as a spicy, sweet bass note, and for that reason bartenders sometimes employ it in homemade infusions. Some fans of Pimm's No. 1, the gin-based liqueur used to make the classic British summer cocktail known as a Pimm's Cup, swear they taste fenugreek in its mysterious—and highly secret—spice blend.

PIMM'S CUP

1 part Pimm's No. 1
3 parts lemonade
Sliced cucumbers, oranges, and strawberries
Spearmint leaves
Borage blossoms or leaves (optional)

Fill a pitcher or glass with ice and add all the ingredients. Stir well. Borage leaves and blossoms are a traditional garnish but not always easy to find unless you grow them yourself.

GALANGAL

Alpinia officinarum
ZINGIBERACEAE (GINGER FAMILY)

The sharp, spicy flavor of this ginger relative has been popular in Chinese, Thai, and Indian cooking for centuries. Its traditional use as a digestive treatment led to its inclusion in early medicinal tonics that later became popular liqueurs. Today it is still found in some vermouths and bitters, and in eastern European herbal liqueurs such as Liqueur Herbert.

As with other gingers, the rhizomes are used in the spice trade. The plant is allowed to grow for four to six years, reaching about eight feet in height, forming a clump of tall stems topped by strappy leaves. The entire root base can be harvested at once, or just a few rhizomes can be dug out around the edges.

Although several related plants are referred to as galangal, the so-called lesser galangal, *Alpinia officinarum,* is the species recognized as a safe ingredient by the FDA. Other species include greater galangal, *A. galangal,* and *Kaempferia galangal,* sometimes called resurrection lily. All three grow in tropical climates and produce pink and white blossoms that resemble a spray of orchids or tuberose.

GENTIAN

Gentiana lutea

GENTIANACEAE (GENTIAN FAMILY)

Without this tall yellow flower that grows wild in French alpine meadows, any number of classic cocktails would not exist. The Manhattan, the Negroni, and the Old-Fashioned all rely on the bitterness of gentian. Angostura bitters, a staple ingredient found in even the most poorly stocked bars, contains gentian and even broadcasts that fact on the label. Many of the most famous European *amaro*s and liqueurs set aside their secrecy and plainly claim gentian as a key ingredient. Campari, Aperol, Suze, Amaro Averna, and the aptly named Gentiane are just a few of the hundreds of spirits that depend on this plant for bitterness.

Its medicinal use dates back at least three thousand years. Egyptian papyrus from 1200 BC documents its use as a medicine, and it has been continuously used since then. Pliny the Elder wrote that gentian owes its name to King Gentius, ruler from 181 to 168 BC of a Roman province that is now part of Albania.

Gentians are not easy to cultivate. Each species prefers a very specific climate and soil type; many detest rich, loamy garden soil and resist transplanting. Of the over three hundred species identified, only a dozen or so actually do well in gardens. Yellow gentian in particular prefers alpine meadows to farmland; it is protected in parts of Europe and wild harvesting is subject to strict controls. (A poisonous look-alike, *Veratrum album,* also makes foraging for gentian a dangerous pursuit for amateurs.)

One of the reasons wild gentian requires protection is that the root is used in liqueurs and medicine; there is no way to harvest it without digging up the entire plant. The bitter compounds include gentiopicroside and amarogentin, which modern researchers have investigated for their ability to promote salivation and the

DR. STRUWE'S SUZE AND SODA

Dr. Lena Struwe, a botanist at Rutgers University, has made gentian her life's work. She studies the taxonomy, biodiversity, and medicinal uses of the plant—and she collects vintage bottles and posters that feature the plant. This is her favorite gentian-based cocktail.

2 ounces Suze
2 to 4 ounces soda or tonic water
Lemon twist

Pour the Suze over ice, top with soda water to taste, and add a twist of lemon. Santé!

production of digestive juices. (No wonder it is an ingredient in so many aperitifs.) Gentian even has benefits for people undergoing cancer treatments who have difficulty tasting or swallowing food, and it is under investigation as an antimalarial and antifungal drug.

The plant is typically harvested at four or five years of age, when the long, tuberous roots weigh several pounds. Eight tons are collected every year in the Pyrenees alone; far more is harvested in the Alps and nearby Jura mountains. The bitter components reach their peak in springtime and are more prominent in gentian harvested at higher altitudes, making the precise timing and location of the collection critical.

The bracingly strong bitterness of gentian is precisely what makes it so appealing in a liqueur. It acts as a foil to sugars and floral flavors, giving cocktails like the Negroni the backbone they require. Yellow antioxidants called xanthones give gentian liqueurs a natural golden color; this is evident in products like Suze, a white-wine-based gentian aperitif that is well loved in France but only just becoming available in the United States.

Gentian was also a key ingredient in Moxie, a soda that was once more popular than Coca-Cola. Essayist and *Charlotte's Web* author E. B. White once wrote in a letter, "I can still buy Moxie in a tiny supermarket six miles away. Moxie contains gentian root, which is the path to a good life. This was known in the second century before Christ, and is a boon to me today."

GERMANDER

Teucrium chamaedrys
LAMIACEAE (MINT FAMILY)

This low-growing perennial herb from the Mediterranean is known to gardeners as an edging plant in knot gardens. With a stiff, upright habit and dark, glossy, narrow leaves followed by spikes of small pink blooms, germander is perfect for marching in a straight line through formal landscapes. The leaves give off a strong herbal fragrance similar to sage, a close relative. Medieval physicians prescribed it for a wide range of ailments, and over time it became a bitter flavoring in vermouths, bitters, and liqueurs.

GINGER

Zingiber officinale

ZINGIBERACEAE (GINGER FAMILY)

This tropical plant might not look like much—it rarely blooms, instead producing only green, reedy stalks three to four feet tall with strappy leaves—but its root is one of the world's oldest spices. A native of China and India, ginger was an important part of ancient Chinese medicine and was adopted for medicinal use in Europe after arriving on the earliest trading routes. It has been used to flavor beer since the Middle Ages and adds a note of heat and spiciness to herbal liqueurs, bitters, and vermouth. Domaine de Canton, Snap, and the King's Ginger are just a few modern liqueurs that add a bite of ginger to cocktails.

Today ginger is grown around the world, primarily in Nigeria, India, Thailand, and Indonesia. How the plant is grown, harvested, and stored has tremendous influence on its flavor. Roots harvested after only five to seven months are quite mild, but the flavorful oil content increases quickly after that, peaking at about nine months of age. Plants grown in the shade tend to have more citrus flavor than those grown in the sun. If the root is harvested and dried rather than sold fresh, about 20 percent of the oil simply evaporates, taking with it the bright, citrus characteristics and leaving more zingiberene, the compound that gives it such sharp spiciness. There are several dozen different varieties cultivated for the spice trade today, and each has its own distinct reputation.

Ginger beer was once a mildly alcoholic beverage made with water, sugar, ginger, lemon, and yeast. In its modern nonalcoholic incarnation, also called ginger ale, it plays a starring role in many classic cocktails. A shandy is a mixture of equal parts beer and some fizzy soda like lemonade; a shandygaff is beer and ginger beer. A Dark and Stormy is a mixture of two parts dark rum and three parts ginger beer, served over ice. Gosling's has actually trademarked the name Dark 'n Stormy and recommends, not surprisingly, that you mix it with its brand of dark rum and its brand of ginger beer.

The Moscow Mule, invented in 1941 by a vodka distributor, not only put ginger beer to good use but also introduced Americans to vodka, helping sales of Smirnoff triple in just a few years. It is traditionally served in a copper mug, but this is merely a marketing gimmick. The story goes that a vodka distributor and a bartender concocted this drink to make use of the bartender's unsold ginger beer and to jump-start vodka sales. Apparently the bartender's girlfriend owned a company that manufactured copper mugs, so her product became part of the recipe, too.

MOSCOW MULE

½ lime

1½ ounces vodka

1 teaspoon simple syrup (optional)

1 bottle ginger beer (Try Reed's
    or another natural, not-too-sweet ginger soda)

Fill a copper mug or highball glass with ice. Squeeze the lime over the ice and drop it in the glass. Add the vodka and simple syrup, if desired, and fill the glass with ginger beer.

GRAINS OF PARADISE

Aframomum melegueta

ZINGIBERACEAE (GINGER FAMILY)

The small black seeds of this West African plant deliver a peppery heat, along with a richer, spicier note similar to cardamom and its other relatives in the ginger family. It made its way to Europe through early spice trade routes and became a flavoring not just for food but for beer, whiskey, and brandy as well, sometimes to disguise the flavor of poor-quality or diluted spirits. Today it is still found in some beers (Samuel Adams Summer Ale is a popular example) and remains an important ingredient in aquavit, herbal liqueurs, and gin, including Bombay Sapphire.

Like other plants in the ginger family, it doesn't look like much: the thin, reedy stalks reach just a few feet in height and produce a spray of long, narrow leaves. The purple, trumpet-shaped blooms give way to reddish oblong fruits that each contain sixty to one hundred of the small brown seeds.

The medicinal properties of grains of paradise have helped solve a long-standing problem at zoos. Captive western lowland gorillas often suffer from heart disease; in fact, it is the cause of death for forty percent of them. In the wild, grains of paradise make up 80 to 90 percent of their diet, suggesting that the anti-inflammatory properties of the plant were keeping them healthy. A gorilla health project is under way now to improve the well-being of captive gorillas, with actual grains of paradise—not gin—under consideration as a prescription for better living.

JUNIPER

Juniperus communis
CUPRESSACEAE (CYPRESS FAMILY)

ocktail historians are on a race to discover the earliest precursor to gin in medical literature. Franciscus de le Boë Sylvius, a Dutch physician working in the seventeenth century, had the lead for a while, as he was making use of juniper extracts in medicinal potions. Now the winner seems to be Belgian theologian Thomas van Cantimpré, whose thirteenth-century *Liber de Natura Rerum* was translated to Dutch by a contemporary, Jacob van Maerlant, in his 1266 work *Der Naturen Bloeme*. The text recommended boiling juniper berries in rainwater or wine to treat stomach pain. That's not gin, but anything that combines juniper and alcohol is a step in the right direction.

Which is not to say that the Dutch invented the use of juniper as medicine. The Greek physician Galen, writing in the second century AD, said that juniper berries "cleanse the liver and kidneys, and they evidently thin any thick and viscous juices, and for this reason they are mixed in health medicines." This certainly suggests a mixture of juniper berries and alcohol, although that, too, would have tasted nothing like the superb gins we drink today.

Junipers are members of the ancient cypress family. They made their earliest appearance during the Triassic period, 250 million years ago. This puts them on the earth at a time when most of the land masses were grouped together in a single continent called Pangaea—and explains why a single species, *Juniperus communis,* can be native to Europe, Asia, and North America.

Because junipers have been around so long, several subspecies have evolved. The juniper used most widely in gin is *J. communis communis,* a small tree or shrub that can live for up to two hundred years. They are dioecious, meaning that each tree is either male or female.

PART II

169

HERBS & SPICES

KNOW YOUR GINS

DISTILLED GIN: Alcohol that has been redistilled with juniper and other botanicals, with added flavorings.

GENEVER: A Dutch style of gin distilled from a malted mash similar to that used for whiskey. *Oude* is an older style that is darker in color and has a stronger malt flavor. *Jonge* is a newer style that is lighter in flavor and color, usually owing to more refined distillation techniques. Either may be barrel aged or unaged.

GIN: A high-proof, vodkalike alcohol flavored with juniper and other natural or "nature identical" flavorings.

LONDON GIN or London dry gin: Alcohol that has been redistilled with juniper and other botanicals, with no additional ingredients beyond water or ethyl alcohol.

MAHON: A wine-distilled gin made only on the island of Menorca, off the Mediterranean coast of Spain.

OLD TOM GIN: An old British style of sweetened gin that is making a comeback among classic cocktail aficionados. It was once dispensed in gin palaces, vending-machine-style, from a stylized cat, described this way by British journalist James Greenwood in 1875:
"Old Tom was merely the cognomen of an animal, which on account of its fiery nature and the sharp and lasting effects of its teeth and claws on all who dared to venture on a bout with it, had been selected as being aptly emblematic of the potent liquid called gin."

PLYMOUTH GIN: A type of gin, similar to the London dry style, that can only be made in Plymouth, England.

SLOE GIN: A liqueur produced by macerating sloe berries in gin, bottled at 25 percent ABV or higher.

The pollen from a male shrub can travel on the wind over a hundred miles to reach a female. Once pollinated, the berries—which are actually cones whose scales are so fleshy that they resemble the skin of a fruit—take two to three years to mature. Harvesting them is not easy: a single plant will hold berries in every stage of ripeness, so they have to be picked a few times a year.

Gin distillers prefer juniper berries from Tuscany, Morocco, and eastern Europe. Much of it is still wild-harvested: for example, Albania, Bosnia, and Herzegovina together produce over seven hundred tons of juniper berries per year, much of it gathered in the wild by individual pickers who sell their harvest to a large spice company. The decidedly low-tech method is time consuming: pickers will place a basket or a tarp under a branch, whack it with a stick, and try to dislodge only the ripe, dark blue berries while leaving younger, green fruit alone. Once picked, they are spread out in a cool, dark place to dry. Too much sun or heat would cause them to lose their flavorful essential oils, and a damp environment could invite mold.

The berries contain α-pinene, which imparts a pine or rosemary flavor, as well as myrcene, which is found in cannabis, hops, and wild thyme. Limonene, the lively citrus flavor common in many herbs and spices, is present as well. It is no wonder that juniper is combined with coriander, lemon peel, and other spices to make gin—the same flavor compounds are found in many of those plants, just in different combinations.

COMMON GIN INGREDIENTS

Angelica root	Coriander	Grains of paradise
Bay leaf	Cubeb	Juniper berry
Cardamom	Fennel	Lavender
Citrus peel	Ginger	Orris root

The Dutch were already distilling gin for something other than medicinal use by the time they revolted against Spain, a conflict that began in 1566 and lasted, in one form or another, until 1648. When British soldiers came to the aid of the Dutch, they learned to enjoy a little gin on the battlefield, calling it Dutch courage for the strength it gave the troops. Edmund Waller memorialized it in a 1666 poem called "Instructions to a Painter": "The Dutch their wine, and all their brandy lose / Disarm'd of that from which their courage grows."

Once the English got hold of gin, there was no stopping them. Juniper berries appeared as ingredients in English distillers' recipes in 1639. By the 1700s, unlicensed gin manufacture was legal in England, and crude and quite toxic gin replaced beer as the tipple of choice. A series of reforms led to more licensing and taxation of gin distilleries, and by the nineteenth century England began producing early versions of the excellent crisp, dry gins it is known for today.

Gin is really nothing more than a flavored vodka whose predominant flavor is juniper, so gin drinkers who say they won't drink vodka misunderstand the nature of their addiction. The base spirit itself is generally a mixture of barley, rye, and perhaps wheat or corn. The juniper and other flavorings can be macerated in the alcohol and redistilled, suspended in "botanical trays" in the still or extracted separately and mixed with the finished spirit. Each process extracts different oils from the plants and yields a different result.

Juniper spirit, made by fermenting juniper berries and water to create a juniper "wine" that is then distilled, is sometimes sold as juniper brandy in eastern Europe. St. Nicolaus distillery in Slovakia, for example, sells a juniper brandy as well as a spirit called Jubilejná Borovička that is bottled with a sprig of juniper. It is described as conveying the dubious "pleasure of drinking juniper twig."

Some American distillers are experimenting with local junipers instead of turning to the traditional European sources. Bendistillery in Oregon harvests wild juniper berries for its gin; in fact, the owner says that the reason he started making gin was to put the Pacific Northwest juniper crop to use. Washington Island in Wisconsin is also home to a fine juniper crop; tourists can join juniper picking excursions for Death's Door, a popular local gin distillery.

However, not all junipers are suitable for harvest. The savin juniper (*J. sabina*), the ashe juniper (*J. ashei*), and the redberry juniper (*J. pinchotii*) are just three examples of toxic species; many others have simply not been studied for their potential toxicity. Anyone wishing to experiment with juniper infusions would be well advised to get *J. communis communis* from a reputable source.

Juniper berries are in short supply in England today owing to the loss of wild habitat and a failure to replant older stands. The conservation charity Plantlife UK has launched a campaign to save England's junipers, appealing to the British fondness for gin and tonics as a way to draw attention to their cause and encourage conservation and habitat restoration.

THE CLASSIC MARTINI

The old joke that a martini should be mixed with nothing more than a rumor about vermouth is best ignored. Bartenders who put a splash of vermouth in a glass, swirl it around, and toss it out before filling the glass with gin are not mixing a drink; they're simply selling you a glass of gin. Vermouth is a type of wine, and as long as it is fresh, recently opened, and refrigerated, it is an excellent mixer. A dusty bottle of vermouth opened months ago should be tossed out.

A martini should be a small drink served cold in a small glass. Some bars pour as much as four or five ounces of straight gin in an enormous cocktail glass, leaving drinkers to contend with warm, undiluted gin, which is not a cocktail.

1½ ounces **gin**
½ ounce **dry white vermouth**
Olive or lemon peel

Shake the gin and vermouth vigorously over ice. Strain and pour into a cocktail glass. Garnish with the olive.

LEMON BALM

Melissa officinalis
LAMIACEAE (MINT FAMILY)

While this mint relative does smell strongly of lemon, the most common variety has a citronella note more reminiscent of lemon floor cleaner than anything that might taste good in a cocktail. The cultivar *Melissa officinalis* 'Quedlinburger Niederliegende' has the higher essential oil content that distillers prefer. Those oils include citral and citronellal, linalool, and geraniol, which gives it a slight rose geranium fragrance. The upper leaves and flowers are steam-distilled to extract this potent flavor, which goes into absinthe, vermouth, and herbal liqueurs. It is suspected to be one of the secret ingredients in both Chartreuse and Benedictine.

The genus name *Melissa* comes from the Greek word for "honeybee"; it gets this name because the tiny flowers are so attractive to bees.

LEMON VERBENA

Aloysia triphylla

VERBENACEAE (VERBENA FAMILY)

This wildly fragrant but otherwise unassuming shrub has a dramatic history. It arrived in Europe from its native Argentina in the 1700s but was never properly described in the botanical literature. A botanist named Joseph Dombey collected it again during an ill-fated expedition to Latin America in 1778 but ran into trouble in 1780 when he found himself in the middle of a Peruvian civil war. After surviving the war, a cholera outbreak, and a shipwreck, he made it to Spain in 1785, only to have his collection of rare plant specimens, representing years of work, detained in a customs warehouse until they rotted and died. One of the few plants that survived was lemon verbena. This time, his colleagues took note and the plant was at last properly identified and described.

DOMBEY'S LAST WORD

In honor of Joseph Dombey, a twist on the classic cocktail the Last Word. This version replaces Chartreuse with a more overtly lemon verbena-flavored liqueur and substitutes the lime for lemon. Given the political turmoil that he found himself in, it seems only fitting that this cocktail combines ingredients from three countries that were also in constant upheaval: England, France, and Italy.

½ ounce **gin (Plymouth or another London dry gin)**
½ ounce **Verveine du Velay**
½ ounce **Luxardo maraschino liqueur**
½ ounce **fresh-squeezed lemon juice**
1 sprig **fresh lemon verbena**

Shake all the ingredients except the lemon verbena sprig with ice and strain into a cocktail glass. Rub a lemon verbena leaf around the rim of the glass and garnish with another leaf. If you can't find Verveine du Velay, green Chartreuse is a fine substitute.

GROW your OWN

LEMON VERBENA

Fresh lemon verbena is not the sort of herb generally sold in grocery stores, so it's worth growing yourself if your climate allows it. It is sensitive to cold and dies back to the ground at the first frost. If covered with straw, it can survive temperatures down to about 10 degrees Fahrenheit. Just leave the branches on the plant through the winter, cutting them back in spring when new leaves emerge. Some cold-climate gardeners take a cutting in fall and nurse it through the winter, planting it outside in spring.

Apart from protection from cold, lemon verbena needs very little care. No special fertilizer is required; like many herbs, it actually prefers poor, well-drained soil on the dry side. Plant it in full sun; if it gets any shade at all, it won't be as flavorful. The flavor is extracted from the leaves, which reach peak essential oil content in the fall. It grows to the size of a small tree in frost-free climates; otherwise it reaches eight to ten feet tall in a season and produces flowering stalks covered in tiny white blossoms.

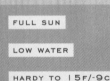

FULL SUN

LOW WATER

HARDY TO 15F/-9C

Unfortunately, Dombey's troubles weren't over. The French government sent him on another expedition to North America, but this time he got as far as the Caribbean island of Guadeloupe before being arrested by the governor, who was still loyal to the monarchy and suspicious of the newly formed French republic that had organized Dombey's expedition. The explorer was able to clear his name, but he was ordered off the island, which suited his purposes anyway. However, his ship was almost immediately captured, probably by privateers working for the British government, and he was again thrown into prison on the nearby island of Montserrat, where he died in 1796.

A shot of verbena liqueur probably wouldn't have offered much consolation to Mr. Dombey, but the herbaceous perennial he helped introduce now lends a sweet, bright lemon flavor to many of the traditional green and yellow liqueurs of southern France and Italy, most notably Verveine du Velay, made by Pagès Védrenne in Le Puy-en-Velay in south-central France. It is also an ingredient in some Italian *amaro*s. On liquor bottles it might be identified as *verveine* in France and *cedrina* in Italy.

LICORICE-FLAVORED HERBS:
A PASTICHE OF PASTIS

LICORICE: THE DRIED ROOT OF A EUROPEAN LEGUMINOUS PLANT WITH PINNATE LEAVES AND SPIKES OF BLUE FLOWERS; AN EXTRACT OF THE ROOT USED IN MEDICINES, LIQUORS, AND CONFECTIONS; A CANDY FLAVORED WITH LICORICE OR A SUBSTITUTE SUCH AS ANISE. ALSO APPLIED TO VARIOUS PLANTS USED AS SUBSTITUTES FOR TRUE LICORICE.

LICORICE: AND NOW FOR A CHEMISTRY LESSON

The licorice flavor in pastis and other such spirits can actually come from several different, and surprisingly unrelated, plants. What each of them have in common is anethole, a licorice-flavor molecule with some unique characteristics. It is soluble in alcohol but not in water, so licorice-flavored drinks are generally higher in alcohol content to keep the anethole molecules from breaking out of solution. But when more water is added—particularly cold water, as is the custom for drinking pastis and absinthe—the anethole separates from the alcohol and forms a milky white or pale green cloud in the drink, called the *louche* in the case of absinthe.

The reason the anethole doesn't simply float to the top in an oily blob when water is added (as, say, olive oil or butter might float to the top of a bowl of soup) is that anethole has what chemists refer to as a low interfacial tension. Imagine two drops of water next to each other. If they get very close together, they will easy merge and become one drop of water. Water drops have a higher level of surface tension and tend to merge readily. On the other hand, picture two soap bubbles. They might stick to each other but will not necessarily merge together to form one larger bubble. That's because they have a lower surface tension. The low surface tension of anethole slows down the speed at which those droplets come together to form one oily mass. That means that pastis or absinthe will stay uniformly cloudy when water is added to the glass, as the anethole breaks loose but resists clumping together.

Some distillers use chill filtration to remove any of these large, unstable molecules that would otherwise cloud the drink in the presence of water or even cold temperatures, which is why some licorice-flavored drinks don't get cloudy. And some oily, plant-based flavor molecules happen to be transparent, which means when they break out of suspension they don't cloud a drink the way anethole does.

⊰ ANISE ⊱

Pimpinella anisum

APIACEAE (CARROT FAMILY)

This small, airy herb, native to the Mediterranean and southwest Asia, looks very similar to its close relatives fennel, parsley, and Queen Anne's lace. The tiny fruits it produces, commonly called aniseed, contain high levels of anethole and are widely used in liqueurs, vermouths, and the yellow Italian aperitif Galliano. Anise is sometimes called burnet saxifrage, although it is neither a burnet (a small plant in the rose family) nor a saxifrage (a low-growing alpine plant that thrives in rocky soil).

⚜ ANISE HYSSOP ⚜

Agastache foeniculum

LAMIACEAE (MINT FAMILY)

In spite of its name and its anise flavor, this native North American mintlike herb does not actually contain significant quantities of anethole. Its flavor comes mostly from estragole, another flavor compound also found in tarragon, basil, anise, star anise, and other herbs. While it can be used by distillers, it is more likely to be used as a mixer. Its name is somewhat misleading, as it is neither an anise nor a hyssop, both plants also used for their licorice flavor.

⚜ FENNEL ⚜

Foeniculum vulgare

APIACEAE (CARROT FAMILY)

This tall, striking perennial herb with fine, lacy foliage and bright yellow flowers is used in a variety of cuisines throughout the Mediterranean, North Africa, and Asia. The bulb, leaves, and stalks are all edible, but it is the fruit—often called a seed, although the seeds are actually found inside the tiny oblong fruits—that is used to flavor absinthe, pastis, and other liqueurs.

A cultivar called Florence fennel (*Foeniculum vulgare* var. *azoricum*) is grown more for its bulb, but it also produces seeds with higher levels of anethole and limonene, giving it a sweet, lemony flavor. Another variety, sweet fennel (*F. vulgare* var. *dulce*), also has higher levels of these flavors and is used for essential oil production and distillation. The dulce variety has the added advantage of having very low levels of eucalyptol, which would give an unpleasant medicinal, camphor flavor to spirits. Fennel pollen is also high in these oils, although difficult to collect in any significant quantity.

THE PERFECT PASTIS

1 plane ticket to Paris
1 summer afternoon
1 sidewalk café

Upon arrival in Paris, locate a café that appears to be frequented
by actual Parisians. Secure a seat and order un pastis, s'il vous plaît.
If it is served neat with a jug of cold water, you are expected to mix it
yourself, drizzling the water in until you have achieved a satisfactory
ratio—usually 3 to 5 parts water to 1 part pastis.

⊰ HYSSOP ⊱

Hyssopus officinalis
LAMIACEAE (MINT FAMILY)

This blue- or pink-flowered mint, also called herb hyssop, is native to the Mediterranean and is an ingredient in absinthe, herbal liqueurs, and natural cough medicines. In spite of its popularity in licorice-flavored liqueurs, chemical analysis shows that it actually has more camphor- and pine-flavored components. Extracts can cause seizures in large quantities but are considered safe in the kinds of low doses used in spirits.

⊱ LICORICE ⊰

Glycyrrhiza glabra

FABACEAE (BEAN FAMILY)

This small, southern European perennial is actually a type of bean, but unlike most beans, it reaches only two or three feet tall and doesn't form a vine. The root is the part of the plant harvested for its flavor. In addition to anethole, it contains high levels of the natural sweetener glycyrrhizin, which can cause high blood pressure and other dangerous conditions in large quantities. Licorice is used in cigarettes to mask harsh flavors and retain moisture, and in candies and liqueurs.

A WORLD *of* LICORICE-FLAVORED DRINKS

ABSINTHE	France
AGUARDIENTE	Colombia
ANESONE	Italy
ANIS	Spain, Mexico
ANIS ESCARCHADO	Portugal
ANISETTE	France, Italy, Spain, Portugal
ARAK	Lebanon, Middle East
HERBSAINT	United States
MISTRA	Greece
OUZO	Greece, Cypress
PASTIS	France
PATXARAN	Spain
RAKI	Turkey, Balkan states
SAMBUCA	Italy

✛ STAR ANISE ✛

Illicium verum

SCHISANDRACEAE (STAR-VINE FAMILY)

Star anise is the fruit of a small Chinese evergreen tree related to the magnolia. The star-shaped fruits, which are harvested while unripe and allowed to dry in the sun, form five to ten points, each containing a single seed. The oils are concentrated not in the seeds themselves but in the star-shaped shell, called a pericarp. The oil is easier and less expensive to extract from star anise than it is from anise, so star anise is more widely used in pastis and herbal liqueurs. In recent years, however, as much as 90 percent of the world's harvest of star anise has been purchased by the pharmaceutical industry to make Tamiflu, a drug used to combat flu pandemics.

The trees grow in China, Vietnam, and Japan. Japanese star anise (*Illicium anisatum*), a close relative, is severely toxic and has poisoned people who picked it by mistake, so it would be unwise to harvest this one in the wild.

⊰ SWEET CICELY ⊱

Myrrhis odorata
APIACEAE (CARROT FAMILY)

Cicely leaves and stems contain enough anethole to make them a useful licorice-flavored ingredient in aquavit and other spirits. Like other members of the carrot family, it is a feathery-leaved perennial with white umbel-shaped flowers. Although it is sometimes called British myrrh, it is not to be confused with the tree called myrrh from which a powerful resin is extracted.

SAZERAC

This classic New Orleans cocktail is the perfect gateway drink for anyone unaccustomed to licorice-flavored cocktails.

1 sugar cube
2 to 3 dashes Peychaud's bitters
1½ ounces Sazerac rye or another rye whiskey
¼ ounce Herbsaint, absinthe, or pastis
Lemon peel

This drink requires a somewhat showy technique, but it's worth learning: Fill an Old-Fashioned glass with ice to get it cold. In a second Old-Fashioned glass, muddle the sugar cube and bitters, and then add the rye. Pick up the first glass, toss the ice into the sink, then swirl the Herbsaint around the glass and toss it out as well. Pour the rye mixture into the Herbsaint-coated glass and garnish with lemon peel.

MAIDENHAIR FERN

Adiantum capillus-veneris
PTERIDACEAE (FERN FAMILY)

With its delicate fan-shaped leaves and dramatic black stems, the maidenhair fern has been a prized conservatory plant since Victorian times. A cosmopolitan species native to North and South America, Europe, and parts of Asia and Africa, this fern has been around long enough to find its way into traditional medicines. One such product, capillaire, made the transition from medicinal tonic to cocktail ingredient.

Seventeenth-century herbalist Nicholas Culpeper recommended capillaire syrup to treat coughs, jaundice, and kidney problems. Over time, the fern became less important as an ingredient, and the term *capillaire* came to refer simply to a syrup of sugar, water, egg whites, and orange flower water. Today the syrup is making a comeback in re-creations of vintage cocktails and punches like the classic Jerry Thomas' Regent's Punch.

CAPILLAIRE SYRUP

Several stems of fresh maidenhair fern
2 cups water
1 ounce orange flower water
1½ cups sugar

Bring the water to a boil, and pour it on the fern. Let stand for 30 minutes. Strain and add the orange flower water and sugar. Reheat, if necessary, to dissolve the sugar. It will keep for a few weeks in the refrigerator and longer in the freezer.

THIS SYRUP COULD BE USED IN ANY RECIPE THAT CALLS FOR SIMPLE SYRUP, BUT A HISTORICALLY ACCURATE EXPERIMENT COULD BE CONDUCTED WITH THE FOLLOWING RECIPE FROM JERRY THOMAS' FAMOUS 1862 MANUAL, *THE BAR-TENDER'S GUIDE.*

JERRY THOMAS' REGENT'S PUNCH

1½ pints strong green tea
1½ pints lemon juice
1½ pints capillaire syrup
1 pint rum
1 pint brandy

1 pint arrack (see note)
1 pint curaçao
1 bottle Champagne
Slice of pineapple

Combine all the ingredients in a punch bowl. The lemon juice can be a bit overwhelming in this original recipe; try scaling it back and using sweeter Meyer lemons instead. It is also improved by topping off each glass with a little extra Champagne. SERVES 30

NOTE: ARRACK IS A BROAD TERM FOR SPIRITS MADE BY DISTILLING THE SUGARY SAP OF COCONUT OR PALM. IT'S NOT EASY TO FIND, BUT BATAVIA ARRACK, MADE FROM SUGARCANE AND RED RICE, IS WIDELY DISTRIBUTED. WHILE THE FLAVOR MAY BE QUITE DIFFERENT, BATAVIA ARRACK IS NONETHELESS A FINE INGREDIENT FOR THIS AND OTHER PUNCHES.

VARIATION: *Substitute "ounce" for "pint" in the above recipe to make cocktails for two. Use about 4 ounces of champagne.*

While maidenhair ferns are widely regarded as nontoxic and are included on the FDA's list of approved food additives, many other species of ferns are poisonous and can cause severe gastrointestinal problems. Some species of ferns, including bracken ferns, also contain carcinogens. In addition, maidenhair ferns are known for their remarkable ability to take up toxins such as arsenic from the soil and therefore should not be gathered in the wild where the soil condition is unknown. For all these reasons, homemade capillaire should be undertaken with care.

MEADOWSWEET

Filipendula ulmaria
ROSACEAE (ROSE FAMILY)

This weedy, swamp-loving perennial forms a dense carpet of leaves topped by two-to-three-foot tall spikes of creamy white flowers. It is native to Europe and parts of Asia, where it has been an ingredient in medicinal tonics since at least medieval times. In fact, the plant contains high levels of salicylic acid, which made it an important ingredient in early aspirin formulations.

As a flavoring agent, meadowsweet gives off a lovely light mixture of wintergreen and almond flavors. Archeological evidence shows that it has been used with other herbs to flavor beer since about 3000 BC; more recently, it has been an ingredient in gins, vermouths, and liqueurs.

NUTMEG AND MACE

Myristica fragrans

MYRISTICACEAE (NUTMEG FAMILY)

The Dutch had a devious strategy for seizing control of the world's nutmeg supply. They realized that the Banda Islands in Indonesia were run by local chieftains who had a long history of competing with one another to sell spices to Arab traders. The Dutch offered each chieftain a treaty guaranteeing protection against hostile competing tribes in exchange for a monopoly on their goods—primarily nutmeg. When the treaties proved difficult to enforce, the Dutch massacred most of the islanders and enslaved the rest. Soon the islands were converted to nutmeg plantations entirely under Dutch control.

They held on to their monopoly through the eighteenth century, even going so far as to burn a warehouse filled with nutmeg in 1760 to hold down supply and boost prices. By the early 1800s, French and British traders managed to smuggle saplings off the islands and establish plantations in French Guiana and India, where most nutmeg is produced today.

The tree that was the object of such intense scheming and warfare is a graceful evergreen that reaches over forty feet in height and produces a fruit that looks like an apricot. The pit—the seed inside that fruit—is what we know as nutmeg. Surrounding that seed is a lacy red covering called an aril. In the spice trade, it's known as mace.

Mace has a stronger, more bitter flavor and is lighter in color, but it is more expensive: only one pound of mace can be extracted from a hundred pounds of nutmeg. The aromatic compounds dissipate so quickly that it should be ground fresh.

Nutmeg is a key ingredient in spicy liqueurs; it is especially evident in Benedictine. It is delicious grated fresh into autumn cocktails made with apple brandy or rum.

ORRIS

Iris pallida

IRIDACEAE (IRIS FAMILY)

The pharmacy and perfumery of Santa Maria Novella, established by Dominican friars in Florence in 1221, gained notoriety for its use of the rhizomes of iris. They were not the first—Greek and Roman writings mention it—but their perfumes, cordials, and powders contained liberal doses of this rare and precious substance.

Orris was popular not so much for its fragrance—although it does contain a compound called irone that gives it a faint violet smell—but as a fixative, holding other fragrances or flavors in place by contributing a missing atom that would otherwise make the fragrance volatile and easily released from the solution it is suspended in.

None of this chemistry was understood at first. Perfumers and distillers would also not have understood why the rhizomes had to dry for two to three years before they become effective as a fixative. We now know that it takes that long for a slow oxidation process to occur, bringing about the chemical change that causes irone to form from other organic compounds present in the rhizome.

Only about 173 acres of orris are cultivated worldwide; most of the orris is either or *I. pallida* 'Dalmatica', grown in Italy, or its descendant *I. germanica* var. *Florentina,* grown in Morocco, China, and India. *I. germanica* 'Albicans' is also used in orris production.

To extract the orris, the rhizome must first be pulverized and steam-distilled to produce a waxy substance called orris butter, or *beurre d'iris.* Then alcohol is used to extract an absolute, which is a perfumer's term for a stronger version of an essential oil.

Orris is found in nearly every gin and in many other spirits. Its popularity in perfume is due to the fact that it not only holds the fragrance in place but clings to the skin as well. It also happens to be a very common allergen, which explains why allergy sufferers might be sensitive to cosmetics and other fragrances—as well as gin.

PINK PEPPERCORN

Schinus molle

ANACARDIACEAE (CASHEW FAMILY)

This fruit comes from the Peruvian pepper tree, a member of the most interesting of plant families. Within Anacardiaceae one finds mangoes, cashews, shellac—and poison ivy, poison sumac, and poison oak. It is, therefore, a family that should be approached with some caution: people who are highly sensitive to poison ivy, for instance, may find that mango rind gives them a rash. Fortunately, mango flesh is perfectly safe, as is the cashew nut itself, minus the shell. And while *Schinus molle,* which grows throughout warmer regions in the United States, is a safe spice, its relative *S. terebinthifolius,* which is found throughout Latin America, can cause a dangerous reaction. (They are easy to tell apart: *S. molle* has long, narrow leaves and *S. terebinthifolius* has glossy oval-shaped leaves.)

The pink peppercorn's history as a drink ingredient begins around 1000 AD, at the remarkable Cerro Baúl brewery in ancient Peru. Archaeological evidence shows that the Wari people settled the area around 600 AD and set up facilities to make corn-based beer flavored with the peppercorns. Women held the high honor of brewmaster. The Wari burned their brewery in 1000 AD—perhaps fleeing the area during warfare—but early Spanish friars reported the use of peppercorn to make wine several centuries later, suggesting that their traditions survived. Today it is used as a flavoring in beer, gin, flavored vodka, and bitters.

SARSAPARILLA

Smilax regelii

SMILACACEAE (GREENBRIAR FAMILY)

Many people know sarsaparilla as an old-time soda similar to root beer. In fact, the drink called sarsaparilla was made with sassafras, birch bark, and other flavors, but no actual sarsaparilla. The climbing, thorny vine that really is sarsaparilla has been used as traditional medicine in its native Central America and was even championed once as a cure for syphilis. It also played a key role in the development of birth control pills: in 1938, a chemist named Russell Marker discovered that a plant steroid derived from sarsaparilla could be chemically altered to make progesterone. This process was too expensive to implement on a large scale, so he found an easier plant to work with: a wild yam from Mexico. His discoveries helped launch the birth control pill and the sexual revolution that followed. (It also launched rumors that sarsaparilla contained natural testosterone and increased sexual potency, none of which is true.)

The ground, dried root is available from spice suppliers and can be used as an ingredient in liqueurs and other spirits—but the ground root of another vine, Indian sarsaparilla (*Hemidesmus indicus*) is also popular in the spice trade for its sweet, spicy, vanilla flavor. Oregon's Aviation gin relies on Indian sarsaparilla for a rich, deep cola flavor that its distillers believe helps the high notes stand out and makes Aviation distinctive.

SASSAFRAS

Sassafras albidum

LAURACEAE (LAUREL FAMILY)

I magine the situation that European colonists found themselves in when they arrived in North America. They brought what food and medicine they could, but much of it was already consumed, or spoiled, by the time they came ashore. They encountered plants and animals they'd never seen before and had no choice but to undertake a dangerous game of trial and error to find out what they could eat or drink. Any berry, leaf, or root could either save them or kill them.

One such plant was sassafras, a small and highly aromatic tree native to the East Coast. The leaves and root bark were put to use as a medical remedy right away: in 1773, sassafras was described in an early history of the colonies as being used "to promote perspiration, to attenuate thick and viscous humours, to remove obstructions, to cure the gout and the palsy." Godfrey's Cordial, a popular nineteenth-century cure-all, included molasses, sassafras oil, and laudanum, a tincture of opium.

Filé, or ground sassafras leaves, became a key ingredient in gumbo. The root bark was used in tea and in early sarsaparilla and root beer, which would have had a very low alcohol content or none at all. It was a classic American spice. However, in 1960, the FDA banned the ingredient because a major constituent of the plant, safrole, was found to be carcinogenic and toxic to the liver. Today it can only be used as a food additive if the safrole is extracted first. Fortunately, the leaves contain much lower levels of safrole, so filé remains available to Cajun cooks.

A Pennsylvania company called Art in the Age of Mechanical Reproduction has resurrected the traditional recipe for sassafras-based brews in the form of Root liqueur, a rich root-beer-flavored spirit that contains birch bark, black tea, and spices—but no sassafras. Instead, a mixture of citrus, spearmint, and wintergreen has been substituted, but the flavor is overwhelmingly true to the sassafras tree.

SUNDEW

Drosera rotundifolia
DROSERACEAE (SUNDEW FAMILY)

Carnivorous plants don't often find their way into cocktails—or at least, they haven't so far. If bourbon can be infused with bacon and stinging nettles can flavor simple syrup, perhaps insect-eating bog plants are poised for a comeback on drink menus.

A tiny carnivore called a sundew was once used in cordials. It is native to Europe, the Americas, and parts of Russia and Asia, where it thrives in swamps during the summer and curls up to wait out the long, cold winter. The sundew, a tiny rosette of narrow red leaves, earns its living by luring insects with a sweet, sticky nectar, then using digestive enzymes to suck nutrients from its victims.

The cordial made from the plant was called rosolio, a term that is now used to refer to any liqueur consisting of fruit and spices steeped in a spirit, sometimes mixed with wine. Scholars disagree over the origin of the word *rosolio* (some believe it actually refers to an infusion of rose petals in alcohol), but it might have come from an early word for sundew, *rosa-solis.* Sir Hugh Plat, writing in 1600, offered a recipe for rosolio that clearly referred to the carnivorous plant, as he even recommended picking out the bugs before infusing it, a step modern bartenders would be well advised to follow: "Take of the hearbe Rosa-Solis, gathered in Julie one gallon, pick out all the black motes from the leaves, dates halfe a pound, Cinnamon, Ginger, Cloves of each one ounce, grains halfe an ounce, fine sugar a pound and a halfe, red rose leaves, greene or dried foure handfuls, steepe all these in a gallon of good Aqua Composita in a glasse close stopped with wax, during twenty dayes, shake it well together once everie two dayes."

Although sundew rarely makes an appearance behind the bar today, there is a German liqueur, Sonnentau Likör, that claims it as an ingredient. Collecting a sufficient quantity of sundew from marshes, and picking the bugs from it, might be more effort than the average cocktailian wishes to undertake, but it's probably a safe enterprise. Sundew has no known toxicity and has even shown limited promise as a cough treatment and anti-inflammatory—demonstrating, once again, that those medieval herbalists might have known what they were doing.

SWEET WOODRUFF

Galium odoratum
RUBIACEAE (MADDER FAMILY)

This low-growing perennial puts out beautiful star-shaped leaves and, in spring, even smaller white, star-shaped flowers. Although it could easily be overlooked as an insignificant, shade-loving woodland groundcover, it gives off a sweet, grassy fragrance, which is an indicator that it contains high levels of potentially toxic coumarin. For this reason, the plant is not considered a safe food additive in the United States—except as a flavoring in alcoholic beverages.

Sweet woodruff is a traditional ingredient in May wine (or *Maiwein*), a German aromatized wine made by infusing the wine with sprigs of woodruff in early spring, before the coumarin levels in the plant have risen to a dangerous level. It is often served with fruit at May Day festivals.

TOBACCO

Nicotiana tabacum

SOLANACEAE (NIGHTSHADE FAMILY)

Smokers may insist that nothing goes better with a drink than a cigarette—but combining them in the bottle? Tobacco liqueur is a strange concoction that could only have been invented in the Americas. In anthropologist Claude Lévi-Strauss's 1973 book *From Honey to Ashes*, he described the practice of soaking tobacco in honey in Colombia, Venezuela, and Brazil. Because fermented honey drinks were also known in South America, it is not inconceivable that people were drinking tobacco in a fermented form.

Native Americans have been cultivating and smoking tobacco leaves for over two thousand years, but Europeans had never heard of the plant—and, in fact, had not smoked much of anything—until explorers brought it back from the New World. It didn't take long for the plant to spread to India, Asia, and the Middle East. At first it was embraced as a kind of medicine: people thought it would treat migraines, ward off the plague, subdue coughs, and cure cancer.

The plant's active ingredient, a neurotoxin called nicotine, is meant to kill insects, but it also happens to kill humans as well. Something called tobacco liquor was widely recommended as a bug spray in the nineteenth century—but it had little to do with the tobacco liqueurs that have been introduced recently.

The best known of these liqueurs is Perique Liqueur de Tabac, distilled at the Combier facility in France through a process that, according to the distillers, leaves no detectable trace of nicotine in the bottle. (Nicotine has such a high boiling point—475 degrees Fahrenheit—that it might not rise through the still at all.) Made with a grape eau-de-vie spirit and aged in oak for over a year, this liqueur is sweet, aromatic, and decidedly different. It comes from a particularly strong and flavorful tobacco strain that is only found in St. James Parish, Louisiana.

Perique tobacco was probably grown by native people for at least a thousand years in that region; settlers have been cultivating and processing it for just two hundred years. The leaves themselves are processed in a way that any distiller would appreciate: they are slightly dried, bundled, then packed into whiskey barrels, where the remaining juice slowly ferments. This adds earthy, woodsy, and fruity flavors to the finished tobacco. In fact, one study identified 330 flavor compounds, 48 of which were previously unknown in tobacco. The tobacco has experienced a resurgence as the interest in artisanal and heirloom ingredients extends to smoking; it is sold in high-end blended pipe tobaccos.

The Perique liqueur does not have the strong, toasted tobacco flavor that a good Scotch has. The best way to describe it is to say that it tastes the way sweet, damp pipe tobacco smells. It is the only widely available liqueur of its kind. Historias y Sabores, a distillery in Mendoza, Argentina, makes a tobacco liqueur; apart from that, the most common use of tobacco in cocktails comes from house-made cigar bitters, an infusion of tobacco and spices in a high-proof spirit that appears on upscale bar menus. Such experiments can be dangerous for bartenders to undertake, however. Without scientific monitoring of the sort not normally practiced in bars, an inadvertently high dose of nicotine in a drink could be delivered to customers.

TONKA BEAN

Dipteryx odorata
FABACEAE (BEAN FAMILY)

This tropical tree, native to damp soils along the Orinoco River in Venezuela, produces a sweet and warmly spicy bean. European plant explorers saw the potential in the bean and brought it back to London's Kew Gardens to be cultivated in tropical hothouses. With notes of vanilla, cinnamon, and almond, it was useful as a perfume ingredient and a baking spice. It was also used to cover the nasty odor of iodoform, an early antiseptic, and until quite recently, it was added to tobacco. Chewing tobacco in particular would be sprayed with a solution made from soaking the beans in alcohol.

It was inevitable that such a tasty bean would turn up in bitters and liqueurs. One brand, Abbott's Bitters, may have derived some of its flavor from tonka beans, according to chemical analysis of old bottles. It is also rumored to be an ingredient in Rumona, a Jamaican rum-based liqueur. But in 1954, the FDA banned tonka as a food ingredient because it contains high levels of coumarin. Liquor containing tonka bean disappeared, but it took a few more decades for tonka to be eliminated from tobacco products, owing in part to the fact that tobacco companies were not required to disclose their ingredients. It is still found as an adulterant in imitation Mexican vanilla, which is why the FDA advises tourists not to bring the product home from vacation.

Tonka bean has made a comeback of sorts. Europeans can find it in the Dutch Van Wees Tonka Bean Spirit, the German liqueur Michelberger 35%, and the French pastis Henri Bardouin. It's used, sometimes on the sly, by chefs and bartenders who believe that the minute dose of coumarin delivered by a fine grating of the spice over a drink or dessert could not possibly be harmful—and in fact, they argue, cassia cinnamon also contains high levels of coumarin and faces no such restrictions. The flat, wrinkled black beans, resembling large raisins, have become a kind of culinary and cocktail contraband.

VANILLA

Vanilla planifolia
ORCHIDACEAE (ORCHID FAMILY)

When Spanish explorers first tasted vanilla, they might not have realized what a rare spice they'd encountered. The vanilla bean is the fruit of a species of orchid native to southeastern Mexico, and it is unusually difficult to cultivate. Like most orchids, it is an epiphyte, meaning that its roots need to be exposed to air, not soil. It climbs the trunks of trees, thriving in limbs a hundred feet aboveground, and unfurls just one flower per day over a two-month period, awaiting pollination by a single species of tiny stingless bee, *Melipona beecheii*. If the flower is pollinated, a pod develops over the next six to eight months. And although the pods contain thousands of tiny seeds, they are incapable of germinating unless they are in the presence of a particular mycorrhizal fungus.

If that isn't complicated enough, the pods themselves don't taste like much of anything when they are picked. They must first be fermented to activate enzymes that release the vanillin flavor. The traditional way to accomplish this was to dip the pods in water, then spread them in the sun and roll them up in cloth to "sweat" at night. The results were worth the effort: hot chocolate drinks flavored with vanilla were one of the Spaniards' most exciting discoveries.

It is no wonder that the first attempts to transport vanilla orchids back to Europe and grow them in hothouses failed. Until the middle of the nineteenth century, no one knew how to pollinate the plants. Finally a method was developed using a tiny bamboo pick, but even that wasn't easy: as each flower opens for only one day, someone has to be standing by, ready to do the work of the bee. Even today, with most vanilla coming from Madagascar, the flowers must be artificially pollinated because the native bee simply cannot be exported. No wonder it competes with saffron for the title of world's most expensive spice.

Over a hundred volatile compounds have been detected in vanilla, which explains why the flavor of pure extract can be so complex: notes of wood, balsamic, leather, dried fruit, herbs, and spices round out the sweetness of vanillin. This makes it an extraordinarily versatile flavor, useful in perfumes, cooking, and in beverages of all sorts. When Coca-Cola made its ill-fated switch to New Coke, the *Wall Street Journal* reported that the economy of Madagascar nearly collapsed because of the sudden drop in demand for vanilla. The company refused, as always, to comment on its secret formula, but the inference was that the original Coke recipe called for vanilla and the new version did not.

Today the highest-quality vanilla comes from Madagascar and Mexico, although some people prefer the fruitier flavor of Tahitian vanilla. The spice can be found in an impossibly wide range of liqueurs, from spiced citrus spirits to coffee and nut liqueurs to sweet cream and chocolate drinks. Kahlúa, Galliano, and Benedictine are just three examples of products strongly dominated by vanilla flavors.

WORMWOOD

Artemisia absinthium
ASTERACEAE (ASTER FAMILY)

nyone who has never tried absinthe will be surprised to find out that it does not taste at all like *Artemisia absinthium*. Wormwood, a pungent, silvery Mediterranean herb, produces volatile oils and bitter compounds that add a kind of mentholated bitterness to aromatized wines and liqueurs, but it isn't usually the primary flavor. In fact, absinthe tastes more like licorice, thanks to another main ingredient, anise. But wormwood gives it its reputation.

Carl Linnaeus, father of modern taxonomy, gave the plant its Latin name when he published *Species Plantarum* in 1753. The word *absinthe* was already in use to describe the plant, so when Linnaeus named it, he was simply formalizing the traditional name. The drink known as absinthe would start to appear in liquor advertisements just a few decades later. In addition to wormwood and anise, it traditionally contained fennel and perhaps a few other ingredients according to the distiller's preferences: coriander, angelica, juniper, and star anise, for example.

The use of wormwood in wine and spirits dates at least to Egyptian times. It was mentioned in the Ebers Papyrus, an ancient medical text from 1500 BC that might actually be a copy of earlier works dating back several more centuries, where it was recommended to kill roundworms and treat digestive problems. At the same time, in China, medicinal wines were made with wormwood; this has been confirmed through chemical analysis of drinking vessels found at archaeological sites.

People eventually realized that adding wormwood to wine and other distilled spirits actually improved the flavor or at least helped disguise the stench of crude, poorly made alcohol.

DANCING WITH THE GREEN FAIRY

Forget about lighting absinthe-soaked sugar cubes
on fire. The traditional method for drinking absinthe
involves only cold water, with a sugar cube if you like your drinks
sweeter. (Modern, artisanal distillers disapprove of adding sugar.)

The addition of water causes a chemical reaction that releases
flavor and changes the color; this phenomenon is known as
the *louche*, although you might think of it as the arrival of the
green fairy.

1 ounce absinthe
1 sugar cube (optional)
4 ounces ice-cold water mixed with ice cubes

*Pour the absinthe into a clear, fluted glass. Rest a spoon across the
top of the glass. (If possible, use a metal slotted spoon or a traditional absinthe spoon). Place the sugar cube, if desired, on the
spoon. (Try half a sugar cube for less sweetness, or none at all.)*

*Now drip the ice water very slowly over the sugar cube, a few drops
at a time, allowing the cube to slowly dissolve and drip sugary water into the glass. If you're skipping the sugar entirely, simply drip
ice water, one drop at a time, into the glass.*

*The essential oils from the plants are very unstable in the alcohol
solution, so adding cold water breaks the chemical bonds and releases the oils. You'll see the absinthe change to a pale, milky green
as those oils are released—that's the* louche. *Because different
flavor molecules are released at slightly different rates of dilution,
going slowly allows the flavors to emerge one at a time.*

*Continue dripping the water in, as slowly as you can, until you've
mixed one part absinthe to three to four parts ice water. Then drink
it at the same leisurely pace, without going to any extra lengths to
keep it cold. As the drink warms, the flavors continue to emerge.*

GROW
your →
OWN

FULL SUN

LOW WATER

HARDY TO -20F/-29C

WORMWOOD

Anyone who finds absinthe intriguing should try growing a little wormwood—not to drink, as making any sort of decent absinthe requires a still—but simply because it's a beautiful and interesting plant.

Many species are available at garden centers or from mail-order nurseries specializing in herbs. All of them sport exquisite, finely cut leaves. The one you're looking for is rarely labeled wormwood; ask for it by its Latin name instead. The plant can survive winter temperatures as low as –20 degrees Fahrenheit, but prefers a warm Mediterranean climate. Plant it in full sun, but don't worry about giving it rich soil: poor, well-drained, dry soil is all it wants. The plant will eventually reach two to three feet in height and width, but it can get leggy if it isn't pruned. To maintain a well-behaved mound, shear back half the foliage in June.

Wormwood isn't recommended as a cocktail mixer because the flavors are harsh and difficult to manage in a drink. Still, if you're planning on inviting some poets and painters over for an evening of absinthe, cut a few branches and bring them indoors to invoke the spirit of the green fairy.

Like many medicinal tonics, wormwood wine eventually became a recreational drink: vermouth. Wormwood also added a bitter—and antimicrobial—element to beer before the use of hops. And it was put to use in a wide range of Italian and French liqueurs.

Although *A. absinthium* is the best-known species, several other species native to the Alps and collectively referred to as *génépi* are also used in liqueurs, including a liqueur called *génépi* that perhaps best captures the actual flavor of the herb. These tend to be small, rugged plants, some only a few inches tall, that thrive in tough, rocky conditions. The wild species are protected and can only be harvested under very limited conditions.

Rumors of wormwood's dangers are greatly exaggerated: while the plant does contain a compound called thujone that could cause seizures and death at very high doses, the actual amount of thujone that remains in absinthe and liqueurs is actually quite low. The stories of absinthe causing hallucinations and wild behavior among France's bohemian set in the late nineteenth century are mostly false; perhaps this was caused by the extraordinarily high alcohol content of absinthe. It was traditionally bottled at 70 to 80 percent ABV, making it twice as alcoholic as gin or vodka.

Absinthe is legal today in Europe, the United States, and many places around the world. Some governments regulate the amount of thujone that may be present in the finished product—this in spite of the fact that many other culinary plants, including sage, are even higher in thujone and aren't regulated at all.

A FIELD GUIDE *to* ARTEMISIA SPECIES USED IN LIQUEURS

Black *génépi, A. genipi*	White *génépi, A. rupestris*
Glacier wormwood, *A. glacialis*	Wormwood, *A. absinthium*
Roman wormwood, *A. pontica*	Yellow *génépi, A. umbelliformis*
Sagewort, *A. campestris*	

-- moving on to --
FLOWERS

flower:

A COMPLEX ORGAN FOUND IN
ANGIOSPERMS, CONSISTING OF
REPRODUCTIVE ORGANS AND THEIR
ENVELOPES, USUALLY INCLUDING
ONE OR MORE STAMENS OR PISTILS,
A COROLLA, AND A CALYX.

CHAMOMILE

Matricaria chamomilla and *Chamaemelum nobile*

ASTERACEAE (ASTER FAMILY)

T wo different plants in the aster family are called chamomile. Roman chamomile, or *Chamaemelum nobile,* is a low-growing perennial that turns up in lawns, and German chamomile, *Matricaria chamomilla,* is an upright annual. The German type is more widely used as a culinary and medicinal herb. It is also much less likely to provoke allergic reactions, which is a common problem with the Roman species.

The round, yellow center of the flower is actually a composite of many tiny flowers fused together, a common characteristic of sunflowers and other plants in the aster family. The German species is sometimes called *M. recutita;* the word *recutita* or *recutitus* in Latin means "circumcised," suggesting that the rounded head looked familiar to some long-ago botanist. One constituent of German chamomile, chamazulene, imparts a surprising blue-green color to chamomile extracts.

Chamomile flowers contain a rich mixture of aromatic and medicinal compounds that are strongest just after the flowers ripen and dry. In addition to their well-known sedative qualities, pharmacological studies show that the flowers' anti-inflammatory and antiseptic effects actually do help calm the stomach.

The makers of Hendrick's Gin claim chamomile as an ingredient, and a few distillers have made it the central ingredient in their liqueurs. J. Witty Spirits in California makes a chamomile liqueur, and the Italian distillery Marolo infuses chamomile in grappa for a sweet, soothing, and surprisingly floral digestif. It is also a key ingredient in vermouth, and one of the few that vermouth makers will admit to on tours of their facilities.

ELDERFLOWER

Sambucus nigra

CAPRIFOLIACEAE (HONEYSUCKLE FAMILY)

The flowers of the elderberry bush impart a flavor that, until recently, was virtually unknown to American palates. Then, in 2007, a pale yellow liqueur called St-Germain entered the cocktail scene. Although it was marketed as an elegant French liqueur, the taste is probably more familiar to British drinkers, who have been imbibing elderflower wine and nonalcoholic elderflower cordial for years.

The elderberry bush flourishes throughout Europe and the United Kingdom. It is a classic hedgerow plant: it grows wild in the countryside, pushing up new shoots from a massive root base every year. The bush produces tiny purplish black berries that can be pressed into juice, cooked into jam, or made into a homemade fruit wine. Elderberry wine has a robust, fruity flavor that is not to everyone's liking, but unscrupulous wine merchants in the nineteenth century knew they could use it to extend wine and port and no one would know the difference.

IS *SAMBUCA* MADE FROM *SAMBUCUS*?

Sambuca is an anise-flavored Italian liqueur that is lovely all by itself after dinner. (Ignore that nonsense about soaking coffee beans in *sambuca* and lighting it on fire. Just pour a little in a glass after dinner and sip it like an adult.) In addition to its overwhelmingly licorice flavor, it can also get a note of fruity complexity from elderberries. Some black *sambucas* derives their rich midnight purple color from the crushed skins of the elderberry, whereas others use artificial colors.

But it is the flat-topped cluster of honey-scented flowers, not the berries, that contribute their remarkable perfume to elderflower liqueur. No other spirit tastes quite so much like a meadow in bloom; if one tries to imagine what honeybees taste when they dive between a flower's petals, this drink is surely it.

St-Germain's distiller reveals little of its recipe, and what is disclosed has been cloaked in fanciful prose. French farmers, the distiller claims, harvest the blossoms in spring and carry them by "specially rigged bicycles" from the foothills of the French Alps to "local depots." They declare that the flowers are not macerated, but that a secret method allows them to persuade the flavor to reveal itself. The extract is then combined with grape eau-de-vie, sugar, and (although they are vague on this point) probably some citrus fruit. The result is a liqueur that tastes of flowers, honey, and the distant, spectral seduction of fruit—pear, perhaps, or melon.

ELDERFLOWER CORDIAL

4 cups water
4 cups sugar
30 clusters fresh (not brown or decaying) elderflowers *(S. nigra)*
2 lemons, sliced
2 oranges, sliced
1¾ ounces citric acid (available at health food stores)

Bring the water and sugar to a boil and allow to cool. While it is cooling, go outside and cut fresh elderflowers, preferably on a warm afternoon when the fragrance is strongest, and shake gently to evict any bugs. Bring indoors immediately and use the tines of a fork to separate the flowers from their stems. Combine all ingredients in a large bowl or jug and let it sit for 24 hours, stirring and tasting as necessary. After 24 hours, strain the mixture into clean, sterile Mason jars. Store in the refrigerator for up to a month or longer in the freezer.

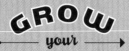

ELDERBERRIES

Although elderberries are used to make jam, wine, and cordial, they can be mildly toxic. All parts of the plant contain a cyanide-producing substance along with other toxins; even the berries should be harvested only when they are completely ripe. North American species of elderberry, including *Sambucus racemosa*, *S. canadensis*, and others, may be more toxic than *S. nigra*, the species that flourishes in English hedgerows. Cooking the berries helps reduce these toxins.

Elderberries tolerate all but the coldest of climates, surviving winter temperatures as low as −30 degrees Fahrenheit. They are shallow-rooted and prefer a topdressing of compost and balanced fertilizer every spring, as well as regular water in summer. To keep a bush fruitful, cut back all canes or branches more than three years old in winter or early spring. Dead or diseased canes should be removed as well. York and Kent are two popular varieties, but find one that works best in your region.

An ornamental variety of *S. nigra*, sold as Black Lace, is popular around the world as a garden plant for its dramatic black foliage and pink flower clusters. It will bloom regardless of whether it has a mate, but to get fruit, there must be another elderberry nearby.

FULL/PART SUN

REGULAR WATER

HARDY TO -30F/-34C

IMBIBING ELDERFLOWERS

Elderflower liqueurs like St-Germain or homemade cordials mix well with almost everything, adding floral and honey notes without ever seeming cloying. Here are a few ways to try it:

• Add a splash to Champagne; float a yellow viola on top.

• Mix a martini with $\frac{1}{2}$ ounce each of elderflower liqueur/cordial and Chartreuse (the green variety if you're brave, the yellow if you're not) in place of the vermouth. Garnish with lemon peel.

• Use soda and elderflower liqueur as a substitute for tonic in a gin and tonic, and add a squeeze of lemon instead of lime.

HOPS

Humulus lupulus and *H. japonicus*

CANNABACEAE (CANNABIS FAMILY)

Beer is not made from hops. It is made from barley, and sometimes other grains, then flavored with hops. But it is impossible to imagine beer without this strange, bitter vine.

Before the discovery, in about 800 AD, that hops could be added to beer to improve its flavor and preserve it, brewers mixed all kinds of strange herbs and spices into their beer. The word *Gruit* is an old German term for the bouquet of herbal ingredients that once went into beer. Yarrow, wormwood, meadowsweet, and even intoxicating and deadly herbs like hemlock, nightshade and henbane all went into the fermentation tank, often with unfortunate results. But that changed once hops migrated from China to Europe in the Middle Ages.

One of the earliest hop farms was established in Bavaria in 736 AD. At that time, brewing and other such scientific and medicinal pursuits were in the hands of monks. Hop farms became common at monasteries throughout Europe and turned up in England by the sixteenth century. With their arrival, a new style of beer was born.

It is difficult to appreciate the storage problems early brewers faced. But just imagine tapping into the last keg in the basement at the end of a long, miserable winter only to find that bacteria had spoiled it months ago. Settlers arriving in the New World on the *Mayflower* may have encountered this very problem: *Mourt's Relation,* an account of the Pilgrims' arrival, suggests that a shortage of beer forced them to make an unplanned landing at Plymouth: "We could not now take time for further search or consideration: our victuals

WHAT'S THE DIFFERENCE BETWEEN ALE AND LAGER?

THAT DEPENDS ON WHO YOU ASK, AND WHEN YOU ASK.
Go back two thousand years to the area that is now Germany and you'll find that *ale* referred to some kind of beerlike fermented drink about which little is definitively known. Go to England in 1000 AD and you'll hear *ale* and *beer* used to refer to two different drinks, *ale* being what we now think of as beer, and *beer* being a drink made with fermented honey and fruit juice.

Then along came hops, and with it the German term *lager*, which was used to differentiate brews that contained hops from those that didn't. Today, however, virtually all beers are flavored with hops. The terms *lager* and *ale* are now used to describe beers brewed with species of bottom-fermenting yeasts or top-fermenting yeasts, respectively. To further confuse matters, Great Britain's commendable Campaign for Real Ale advocates not necessarily for beer made with top-fermenting yeast but for beer made in the traditional English style, which means it's allowed to undergo a secondary fermentation in the cask and is served from a cask at a pub, never bottled.

But where the yeast lived in the fermentation tank is of little concern to the average drinker. It is more important to know that most English beers are called ales, most German and American beers are lagers, and that in bars all around the world, some sort of hand signal will usually get you a brew when words fail.

WHY ARE BEER BOTTLES BROWN?

Brewers learned long ago that dark bottles protect beer from the light and prevent it from developing a skunky "lightstruck" taste. But it wasn't until 2001 that scientists at the University of North Carolina at Chapel Hill found out exactly what causes that nasty flavor. Certain compounds in hops, known as isohumulones, break down into free radicals when exposed to light. Those free radicals are chemically similar to the secretions of skunks. And it doesn't take long for the transformation to happen: some beer drinkers will notice the skunky flavor at the bottom of a pint glass that sat in sunlight while they drank it.

So why are some beers sold in clear bottles? First, it's cheaper. Second, some mass-produced beers are made with a chemically altered hop compound that doesn't break down. But if you see clear-bottled beers sold in a closed box, chances are it's because the brewer knows the taste will degrade quickly in light. And the tradition of adding a wedge of lime to the beer? That's just a marketing ploy to disguise the skunky flavor.

being much spent, especially our beer." With no way to disinfect fresh water, and nothing but salt water all around them, beer might have been the one beverage keeping them alive on their long journey. Once it was gone or ruined, they would have been in real trouble.

But hops came along and turned beer into a far superior product. The hop vine's cones (clusters of female flowers) are laden with yellow glands that secrete lupulin, a resin containing acids that help make beer foamy, give it its bitter taste, and extend its shelf life. These so-called alpha acids are so critical to good beer making that hops are rated according the amount they produce. Aromatic hops are lower in alpha acids but produce delightful flavors and aromas, while bitter hops are higher in alpha acids, make the brew last longer, and contribute more bitterness to counteract the yeasty flavor of the malt.

These vigorous, sturdy vines are closely related to marijuana; there is a vague family resemblance between the dank, sticky cannabis flower buds and the equally sticky and fragrant cones of the female hop vine. Like cannabis, hops are dioecious. Females can grow their highly prized cones without a male around, but they can't set seed and reproduce. Hop farmers select female vines and scour their fields for any uninvited males, which they promptly evict. They don't want the females impregnated, because brewers won't buy seed-infested cones.

Hops can't grow just anywhere: these tall, perennial vines require thirteen hours of sunlight per day while they are growing, and that can only be found in a narrow band around the world at 35 to 55 degrees north and south latitude. That means they are abundant in Germany, England, and elsewhere in Europe. In the United States they are grown mostly in the West: hop farming was pushed westward as powdery and downy mildew diseases made it impossible to grow the vine in eastern states. The hops industry had another advantage in Oregon and Washington: farmers survived Prohibition by shipping dried hops to Asia.

In latitudes 35 to 55 degrees south, hops grow in Australia and New Zealand, and in the north they are grown China and Japan. Attempts have been made to grow them in Zimbabwe and South Africa as

HOP VARIETIES

AROMA (OLD WORLD) HOPS

Cascade
Cluster
East Kent Goldings
Fuggle

Hallertauer
Hersbrucker
Tettnang
Willamette

BITTERING (HIGH ALPHA) HOPS

Amarillo
Brewer's Gold
Bullion
Chinook

Eroica
Nugget
Olympic
Sticklebract

INTERNATIONAL BITTERNESS UNITS: An international scale measuring the level of bitterness contributed by alpha acids in hops.

MASS MARKET AMERICAN BEERS	5 to 9 IBUs
PORTER	20 to 40 IBUs
PILSNER LAGER	30 to 40 IBUs
STOUT	30 to 50 IBUs
INDIA PALE ALE	60 to 80 IBUs
TRIPLE IPAS	90 to 120 IBUs

well, but without the optimal day length, streetlights have had to be installed in hop yards. Meanwhile, optimistic botanists are working on a day-length neutral hop variety that doesn't object to longer or shorter days during its flowering season.

During their growing season, hops are astonishingly vigorous, rising six inches in a single day. The vines stretch away from the central stalk during the day; at night, they wrap themselves around wires or other supports. "You walk through the fields in the late afternoon, and you'll see all these vines reaching out at forty-five degree angles," said Oregon hop farmer Gayle Goschie. "Then you come out the next morning and they're wrapped tightly around the trellis again." They spiral around the trellis in a clockwise direction, which has inspired a couple of botanical urban legends: one is that they grow counterclockwise in the Southern Hemisphere, and the other is that they grow clockwise to follow the sun from east to west. Neither is true. Like left-handedness, they are simply born with a genetic predisposition to grow clockwise, no matter where they are relative to the sun or the equator. (Botanists who study "twining handedness" have discovered that hops are unusual in their proclivity to twine in a clockwise direction; 90 percent of all climbing plants prefer to go counterclockwise.)

Hops don't just climb wire trellises. Tiny barbs on the vines make it possible for them to climb trees or other plants as well. Romans thought the vine killed trees by strangulation and named it "little wolf," which explains the origin of the plant's genus, *Lupulus*.

HOP KILN: ALSO CALLED AN OAST HOUSE IN ENGLAND, THESE DISTINCTIVE BARNS WITH CONE-SHAPED TOWERS WERE USED TO DRY HOPS WHEN THEY CAME IN FROM THE FIELDS. THE HOPS WOULD BE SPREAD ON A FRAME SUSPENDED IN THE UPPER PART OF THE TOWER, AND A FIRE WOULD BE LIT UNDERNEATH TO DRY THE HOPS. THEY COULD THEN BE BAGGED AND STORED IN THE BARN.

Farmers are quick to point out that hops are not a very nice vine. Darren Gamache, a Washington grower, knows what it's like to pick them by hand the way his grandparents did. "The vines have these rough little bristles that are very abrasive, and will even leave welts.

HOPS

No beer garden is complete without an ornamental hops vine. Specialty hop nurseries sell brewers' favorites like Cascade and Fuggle, but a good garden center will also carry an ornamental variety bred for good looks over flavor. The golden hop vine Aureus, with its yellow to lime green foliage, is a widely sold ornamental, as is Bianca, a variety with light green foliage that matures to a darker green, creating a lovely contrast.

Plant hops in full sun or part shade in moist, rich soil. They grow best in latitudes of 35 to 55 degrees north and south, and they are hardy to −10 degrees Fahrenheit. The vines die down to the ground in winter; in a mild winter when the frost does not wither them, they should be cut back to encourage more growth. Expect them to reach twenty-five feet by midsummer and to start blooming by the third year. Once cones emerge, the vines get surprisingly heavy, so give them a sturdy trellis to climb.

The flowers are generally ready to harvest in late August or September. They should feel dry and papery to the touch and smell strongly of hops. Squeeze one that seems ripe; if it bounces back into shape, it's ready to be picked. Once harvested, spread them out on a screen, preferably with a fan underneath them, to encourage air circulation while drying.

FULL SUN

REGULAR WATER

HARDY TO -10f/-23c

Especially when it's hot outside, and you've got salty sweat running into open wounds—it's really uncomfortable," he said. "And a lot of people have an allergic reaction to them." Most hops today are picked by machine for this very reason.

The danger is not over after harvest: freshly picked hops heat up the way a compost pile does and have been known to catch fire. When large bales of hops are packed into storage, they can actually spontaneously combust and burn a warehouse down. Fires at hop yards were a frequent occurrence in the Pacific Northwest's early hop-growing days.

Brewers, for the most part, have little idea that hop farmers must endure scratchy vines, fight warehouse fires, and chase love-struck males out of their fields to get the crop to market. Hops generally aren't even recognizable as cones by the time they get to the brewery: they are pressed into pellets and shipped in vacuum-sealed bags. A few brewers use green hops straight from the fields in a seasonal beer around harvest time; to experience freshly picked hops, look in the fall for "fresh hop" or "wet-hopped" brews.

JASMINE

Jasminum officinale

OLEACEAE (OLIVE FAMILY)

Surely the first person who ever smelled jasmine thought to make a drink of it. Who could resist that sweetly intoxicating fragrance? Jasmine does, in fact, appear in early recipes for cordials and liqueurs: Ambrose Cooper's *The Complete Distiller*, published in 1757, includes a recipe for jasmine water that calls for jasmine flowers, citrus, spirits, water, and sugar. Similar recipes abound in eighteenth- and nineteenth-century cookbooks, and a record from the Great London Exposition of 1862 shows that jasmine liqueurs from the Greek Ionian islands were winning awards.

The jasmine most used in perfumes and liqueurs is *Jasminum officinale,* sometimes called poet's jasmine. (Botanists are in a disagreement at present over whether another species called poet's jasmine or Spanish jasmine, *J. grandiflorum,* is really a separate species.) *J. sambac,* also called Arabian jasmine, or pikake, is popular in Hawaiian leis and is also used in Asian jasmine teas and perfumes. (Jasmine tea, by the way, is usually green tea sprayed with jasmine essence, not a tea made of actual flowers.) None of these are common garden jasmines, but collectors of tropical and fragrant plants have no trouble tracking them down.

Jasmine's fragrance comes several interesting compounds, including benzyl acetate and farnesol, both of which impart a sweet flowery fragrance with notes of honey and pear. Linalool, the ever-present citrus and floral aroma, is there as well, and so is phenylacetic acid. This last substance is also found in honey—and its by-product is excreted in urine. Perfume makers know that, owing to genetic differences in how we experience fragrances, about half the people who inhale jasmine will think of honey, and the other half, unfortunately, will think of urine. They're both right.

Jasmine is not a common ingredient in liqueurs today, in part because of its cost: the makers of Joy perfume love to brag that over ten thousand jasmine flowers go into a single ounce. Jacques Cardin makes a jasmine-infused Cognac, and two American distillers, Koval in Chicago and GreenBar Collective in Los Angeles, produce jasmine liqueurs.

OPIUM POPPY

Papaver somniferum
PAPAVERACEAE (POPPY FAMILY)

T his beautiful annual flower, with its enormous petals the texture of crumpled tissue, has been banned around the world because its pods produce a milky sap laden with opium. While the drug has its uses as a painkiller—morphine, codeine, and other opiates are derived from the plant—it can also be used to make heroin, and for that reason the plant is classified as a Schedule II narcotic in the United States. This has not stopped gardeners from growing the plant in violation of the law; it's actually quite common. Only the seeds can be sold legally, since poppy seeds are used in baked goods. This loophole allows garden centers and seed catalogs to sell them as well.

Perhaps the earliest description of an opium cocktail comes from Homer's *Odyssey,* in which an elixir called nepenthe gave Helen of Troy an escape from her sorrows. While opium was not mentioned specifically, many scholars believe that the wine mixed "with an herb that banishes all care, sorrow, and ill humour" must have referred to an opium-laced drink.

Such a potion continued to be used as a medicinal drink and surgical anesthetic through the Victorian era. At that point laudanum, a medicinal tonic of opium steeped in alcohol, was used to control pain and relieve the suffering brought about by a wide range of ailments. To relieve the symptoms of gout, King George IV liked to tip a little laudanum into his brandy—and then a little more, and a little more, as the highly addictive narcotic took its hold on him.

An opium syrup gained respectability in 1895, when Bayer sold it under the name Heroin. The syrup was banned in the 1920s, and opium cocktails became a relic of the past.

YOU'VE BEEN WARNED

In this age of homemade infusions and bitters, it might be tempting to take a walk on the dark side with opium poppies. But the plant is illegal and its by-products quite dangerous. Don't do it.

ROSE

Rosa damascena and *Rosa centifolia*
ROSACEAE (ROSE FAMILY)

R ed Roses do strengthen the Heart, the Stomach, and the Liver, and the retentive Faculty. They mitigate the Pains that arise from Heat, assuage Inflammations, procure Rest and Sleep, stay both Whites and Reds in Women, the Gonorrhea, or Running of the Reins, and Fluxes of the Belly; the Juice of them doth purge and cleanse the Body from Choler and Phlegm." So proclaimed Nicholas Culpeper in his 1652 medical manual, *The English Physician*. He prescribed rose wines, rose cordials, and rose syrups for a long list of alarming ailments.

Roses are ancient plants that first appeared in the fossil record about forty million years ago. The fragrant garden roses we know today have traveled to Europe from China and the Near East in the last few thousand years. The most popular rose for liqueurs, the fragrant damask rose *Rosa damascena,* came from Syria, where it was being distilled for perfume. European botanists brought it into cultivation as a garden rose and to use in their strange medical preparations, but the Middle East remained the center of rose perfume and rose water production.

Damask roses, with romantic names like Comte de Chambord and Panachée de Lyon, tend to be lush, round, open and highly fragrant flowers with tightly packed petals in shades of pink, rose, and white. Cabbage roses, *R. centifolia,* were developed by Dutch botanists in the seventeenth century for their strong fragrance. The light pink Fantin-Latour is one of the best-known varieties of perfumed cabbage roses.

Most early recipes for rose petal liqueurs, like Culpeper's, called for a maceration of aromatic rose petals, sugar, and fruit in brandy.

Rose water, the watery part of the steam distillation of rose petals that is left behind after the essential oil is removed, is a traditional ingredient in Middle Eastern cuisine.

Lately rose water has become popular as a cocktail ingredient, usually sprayed on the surface of a drink. And a few high-quality rose petal liqueurs are produced in Europe and the United States, including the French Distillerie Miclo's fine liqueur of macerated rose petals, and Crispin's Rose Liqueur, made with an apple spirit base in northern California. The Bols liqueur Parfait Amour claims rose petals as one of its ingredients, along with violets, orange peel, almond, and vanilla. Hendrick's Gin includes a damask rose essence added after the distillation process, along with cucumber, to give it a garden bouquet.

A much less showy species, the eglantine or sweet briar rose *R. rubiginosa,* is grown not for its flowers but for the fruit, called rose hips, that remain after the petals have dropped off the plant. Rose hips are a good source of vitamin C and have been used to make teas, syrups, jams, and wine. An eglantine eau-de-vie is made by a few distilleries in Alsace, and a Hungarian brandy called Pálinka is made from them. Rose hip schnapps and liqueurs also turn up; for instance, the Chicago distillery Koval makes a rose hip liqueur.

✥ SAFFRON ✥

Crocus sativus

IRIDACEAE (IRIS FAMILY)

For such an ancient and important spice, saffron is surprisingly difficult to keep alive, much less harvest. The crocus that we know today as saffron is a triploid—meaning that it has three sets of chromosomes instead of two—and it is sterile. It can only reproduce by creating more corms (a bulblike structure), never by setting seed. It is probably a mutant that has been continuously cultivated since about 1500 BC.

Each corm produces just one purple flower during a two-week period in the fall. That flower opens to reveal the precious three-part red stigma we know as threads of saffron. It takes four thousand flowers to gather just an ounce of saffron. Every few years, the corms must be dug up, divided, and replanted to ensure a good harvest. (Although saffron crocus blooms in the autumn, it must never be confused with the autumn crocus, *Colchicum autumnale,* which is highly toxic.)

Saffron is rich in flavor and aroma compounds. Its bitterness comes primarily from picrocrocin, which breaks down after it is harvested and dries to an oil called safranal. This substance is of great interest to scientists, who find that saffron's long use as a medical herb is not without merit. Limited studies show that it may suppress tumors, aid in digestion, and help scavenge free radicals.

In addition to seasoning Indian, Asian, and European dishes, saffron has been a flavoring for beer and spirits for centuries. Archeologist Patrick McGovern believes that saffron was used as a bittering agent in ancient times; he worked with Dogfish Head to create Midas Touch, a drink made of white Muscat, barley, honey, and saffron, based on residue analysis of drinking vessels found in King Midas's tomb.

Today saffron is cultivated in Iran, Greece, Italy, Spain, and France. Worldwide production is estimated at about three hundred tons, and an ounce of saffron sells for about three hundred dollars at retail, although prices vary widely depending on the quality. (Top-quality saffron is the result of better growing conditions and cultivars; it is worth paying extra to get the good stuff.) Its orange pigment comes from a carotenoid component called α-crocin; this imparts a yellow-ish hue to paella and to yellow liqueurs like Strega. It also colors the yellow version of many traditional yellow and green Chartreuse-like liqueurs made in Spain, France, and Italy. The makers of Benedictine disclose very few of their ingredients but admit to an infusion of saffron.

It is a popular rumor that the extraordinarily bitter Fernet Branca derives much of its flavor from saffron and in fact commands three-quarters of the world's saffron supply. This may be nothing more than a tall tale. If annual production of the spirit is 3.85 million cases, as reported in liquor industry trade journals, that would work out to one-sixth of an ounce of saffron per bottle—roughly twenty-five dollars' worth at retail. With a bottle of Fernet retailing for twenty to thirty dollars, it seems unlikely that it would contain such a large and expensive pinch of the spice—even with massive volume discounts.

VIOLET

Viola odorata
VIOLACEAE (VIOLET FAMILY)

The Aviation cocktail is the Chelsea Flower Show in a glass, combining gin, maraschino liqueur, lemon juice, and crème de violette. And until a few years ago, it was impossible to make properly, because crème de violette had disappeared from the shelves.

That changed thanks to the efforts of Eric Seed, owner of Haus Alpenz, an importer of unusual and hard-to-find spirits. His search for an authentic crème de violette took him to Austria, where Destillerie Purkhart was producing limited quantities for special customers—mostly bakers who used it in chocolates and cakes. They select two varieties of *Viola odorata* for their liqueur: Queen Charlotte (or Königin Charlotte) and March.

VIOLET LIQUEURS

CRÈME DE VIOLETTE: For straight violet flavor, this is the real thing: an infusion of violets, sugar, and alcohol, with a lovely deep purple color.

CRÈME YVETTE: A purple liqueur that may or may not contain violets. The version made by Cooper Spirits International (the same people who gave St-Germain to the world) is a blend of cassis, berries, orange peel, honey, and an infusion of violet petals, giving it a very different flavor than crème de violette.

PARFAIT AMOUR: A purple liqueur with a citrus base, like curaçao, blended with vanilla, spices, and roses or violets.

The sweet violet is a flower of a bygone era; a hundred years ago, they were widely grown and sold in nosegays in flower stalls. The blossoms lasted just a day or two in water, and were meant to be worn or carried for just one evening, so their distinctive fragrance could serve as a woman's perfume.

Sweet violets are sometimes called Parma violets, although a Parma violet is more likely a particular variety of a very similar species, *V. alba*. Violets are unrelated to African violets but are a close relative of two garden center staples, Johnny-jump-ups and pansies.

The fragrance—and flavor—of violets is a tricky one. A compound called ionone interferes with scent receptors in the nose and actually makes it impossible to detect the fragrance after a few whiffs. There's also a genetic component to how we taste ionone: some people can't smell or taste it at all, and others get an annoying soapy flavor rather than a floral essence.

THE AVIATION

1½ ounces **gin**
½ ounce **maraschino liqueur**
½ ounce *crème de violette*
½ ounce **fresh lemon juice**
1 **violet blossom**

Shake all the ingredients except the violet blossom over ice and serve in a cocktail glass. Some versions of this recipe call for less crème de violette *or less lemon juice; adjust the proportions to your liking. Garnish with the violet blossom. (A pansy or Johnny-jump-up would be a botanically appropriate substitution.)*

-- continuing straight ahead to --
TREES

tree:

AN UPRIGHT PERENNIAL PLANT
WITH A SELF-SUPPORTING
SINGLE TRUNK OR STEM
MADE UP OF WOODY TISSUE
SHEATHED IN BARK, OFTEN GROWING
TO A SUBSTANTIAL HEIGHT.

ANGOSTURA

Angostura trifoliata
RUTACEAE (RUE FAMILY)

The manufacturers of Angostura bitters spent decades in court defending their right to the product's name—all the while refusing to say whether it was actually made from the bark of the angostura tree. The late-nineteenth- and early twentieth-century battle set legal precedents around the world at a time when trademark law was still a very new idea.

First, let's address the tree itself: the angostura tree has almost as many names as the bitters that have claimed it as an ingredient. Alexander von Humboldt, a German explorer and botanist, described the tree during an expedition to Latin America that lasted from 1799 to 1804. He wanted to call it *Bonplandia trifoliata,* after botanist Aimé Bonpland who accompanied him on the journey. It has also appeared in botanical literature as *Galipea trifoliata, Galipea officinalis, Cusparia trifoliata,* and *Cusparia febrifuga.* The shrublike tree grows wild around the city of Angostura, Venezuela (now called Ciudad Bolívar). It produces dark green leaves arranged in groups of three (this is where it gets the name *trifoliata*), and fruit split into five segments—a bit like citrus fruit, which are also in the rue family—each of which contains one or two large seeds.

While botanists debated its name, pharmacists debated its medicinal qualities. Alexander von Humboldt wrote that an infusion of the bark was used as a "strengthening remedy" among Indians in Venezuela and that monks were sending it back to Europe in hopes that it could be used to fight fever and dysentery. Throughout the nineteenth century, the bark was described in pharmaceutical literature as a tonic and stimulant that could treat fevers and a variety of digestive ailments. Recipes for angostura bitters, a combination of angostura bark, quinine, and spices soaked in rum, were easy to find in medical journals of the day.

THE CHAMPAGNE COCKTAIL

This classic drink is an excellent way to appreciate the flavor of the angostura tree. The Fee Brothers version proudly claims to contain the bark.

1 sugar cube
3 to 4 dashes Fee Brothers Old Fashion Aromatic Bitters
Champagne
Lemon twist

Drop the sugar cube into a flute, splash a few drops of bitters on the sugar, and fill with Champagne. Garnish with a twist of lemon peel.

The brand we know as Angostura began, the company claims, in 1820, when German physician Johannes G. B. Siegert arrived in a city in Venezuela called Angostura. He created a kind of medicinal bitters using local plants, which was sold under the name Aromatic Bitters, listing the place of manufacture as Angostura, Venezuela. In 1846, the name of the city was changed to Ciudad Bolívar in honor of independence leader Simón Bolívar. In 1870, the doctor died, and the sons later moved the company to Trinidad, seeking political stability. Still the "Aromatic Bitters" label bore the name of Dr. Siegert of Angostura, with the company's new location.

By this time, European nations and the United States were starting to pass trademark laws, and the Siegert brothers wanted in on the action. In 1878, they brought suit in the British courts against a competitor selling Angostura bitters, claiming that their own bitters were widely known as Angostura bitters, even though they were not made in Angostura and the words *Angostura bitters* had not actually appeared on their label until *after* their competitor had started using the name.

SODA *and* BITTERS

If you ever find yourself in the unfortunate position of sitting in a bar when you are unwilling or unable to have a drink, order a club soda with bitters. It has the advantage of looking like a proper drink and is surprisingly restorative.

The competitor—a man named Dr. Teodoro Meinhard—employed a brilliant defense. He stated that his bitters were called Angostura bitters because they contained angostura bark. While the manufacturer of a brand of bitters would normally keep the ingredients secret, the law stated that no one could trademark a name that simply reflected the contents of the product. Anyone can call their products by a name like orange juice, chocolate bars, or leather shoes; those names simply and plainly state what the item is. Meinhard wasn't trying to claim the name Angostura bitters for his exclusive use—he was just trying to keep the Siegerts from doing so. His strategy was partially successful: while the judge ruled that his use of the name Angostura bitters was a clear attempt to defraud customers into buying his product rather than Siegert family version, the ruling also stated that the term *Angostura bitters* did not deserve full protection under English law.

The lawsuits continued in the United States. In 1884 a series of legal actions between the Siegert brothers and C. W. Abbott & Co. began for much the same reason. Abbott also claimed that angostura bark was a key ingredient in his bitters, which gave him protection under the law. Once again, the Siegerts remained silent on their own formula, claiming that the name came from the city, not the tree. This time, things did not go so well for the Siegerts.

The judge ruled that no one can claim a monopoly on the name of a city, even if that city's name had been changed decades ago. And no one could trademark the name of an ingredient or another term that simply describes what the product is. Besides, the judge pointed out, the Siegerts weren't using the term *Angostura bitters* at all until their competitors did so. They had been calling their product Aromatic Bitters; it was only the general public that had taken to calling them Angostura bitters.

Continuing his examination of the evidence, the judge chided the Siegerts because their label still read "Prepared by Dr. Siegert" even though the doctor was deceased. The Siegerts lost their case and Abbott continued to sell Angostura bitters. In subsequent rulings, judges found even more to dislike about the Siegerts' case, including their unsubstantiated claim that the bitters had medicinal purposes. Their luck ran out in Germany as well, where their application for a trademark was denied by a judge who flatly stated that angostura bark was used in the preparation of angostura bitters, and as such the name could not be trademarked.

It was not until 1903 that judges finally began ruling in favor of the Siegerts and granting them the exclusive right to the name Angostura bitters. The Abbott company, commenting not just on its own case but on other similar cases that were now being decided in favor of the Siegerts, issued a statement that expressed its frustration: "Our bitters are made out of Angostura bark. This is the point in our case. And the court did not pass upon it."

In February 1905, the United States updated its trademark laws. It took only three months for the Siegert brothers to file their application under the new law. The application claimed that "the trade mark has been continuously used by us and by our predecessors in business for about the last 74 years past," and that "no other person, firm, corporation or association" had the right to use the trademark. It was approved.

Today the label remains more or less unchanged from the original patent application, with a few exceptions. By 1952, the company had filed an updated label design omitting medical claims and a suggestion that bitters be served to children, and a new phrase had been added: "Does Not Contain Angostura Bark."

So was angostura bark ever an ingredient in Dr. Siegert's recipe or only in those of his competitors? The Siegerts managed to get through thirty years of litigation without revealing their secret formula in any published court record. They did claim that their bitters would treat stomachaches and fevers—the very maladies angostura bark was supposed to treat. (They also said that the bitters should not be used "in the manufacture of cocktails" but added that they should be dashed into a wine glass and topped with rum, wine, or any other spirit, and taken "before breakfast or dinner, or at any other hour of the day, if you should feel inclined," which sounds rather like a cocktail. They also recommended applying them to "new rum" to improve the taste.)

Another strange clue as to the original ingredients comes in the form of an advertisement placed in a theater magazine by the Siegert company in 1889. The ad claims that Dr. Siegert met Alexander von Humboldt in Venezuela in 1839 and prescribed his bitters to the explorer when he became sick. There was just one problem with this story: von Humboldt was in Berlin in 1839. He had, in fact, become sick in Venezuela during his 1799 to 1804 expedition and had been treated with angostura bark—the one ingredient that the company now claims not to use in its bitters.

It's hard to believe that a medicinal bitters, invented in Angostura, Venezuela, that claims to treat fever and stomach problems, would not contain a well-known plant that grew in the area and was already used for those very problems. The use of angostura bark in pharmaceutical preparations is well documented in the nineteenth century. In fact, the realization that angostura bark was sometimes adulterated with the poisonous bark of the strychnine tree led to widespread warnings to druggists to be on the alert when compounding their own angostura bitters. Obviously, the bark was widely used at one point. Why would Dr. Siegert have omitted it from his formula?

The new trademark laws passed at the end of that century made one thing clear: anyone who made bitters with angostura bark would be legally entitled to call their product angostura bitters because it was a plain statement of the nature of the product. The only way to trademark the name would be to make the case about something other than the ingredients—and that's what the Siegert brothers did.

If their formula once contained angostura bark, when did they drop the ingredient? It's possible that Dr. Siegert realized early on that the bark could be mistaken for strychnine bark and decided to steer clear of it. Or if he did use it, perhaps the recipe changed when the company moved to Trinidad, or after the Siegerts' legal conundrum became apparent.

Or, as astute readers might have already wondered, perhaps the formula never changed. After all, the label on the bottle today only says that the product does not contain angostura *bark*. No mention is made of another legally recognized ingredient, angostura extract, or of the tree's trunk, leaves, roots, flowers, or seeds.

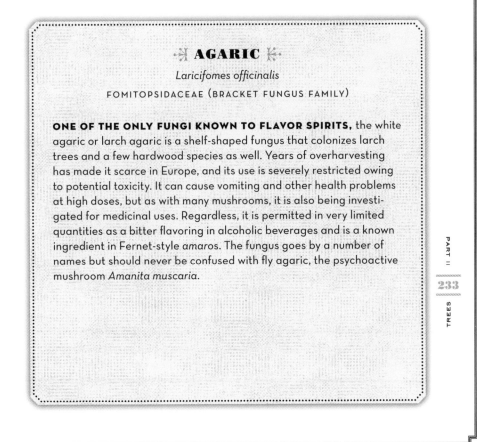

⊹ AGARIC ⊱

Laricifomes officinalis

FOMITOPSIDACEAE (BRACKET FUNGUS FAMILY)

ONE OF THE ONLY FUNGI KNOWN TO FLAVOR SPIRITS, the white agaric or larch agaric is a shelf-shaped fungus that colonizes larch trees and a few hardwood species as well. Years of overharvesting has made it scarce in Europe, and its use is severely restricted owing to potential toxicity. It can cause vomiting and other health problems at high doses, but as with many mushrooms, it is also being investigated for medicinal uses. Regardless, it is permitted in very limited quantities as a bitter flavoring in alcoholic beverages and is a known ingredient in Fernet-style *amaros*. The fungus goes by a number of names but should never be confused with fly agaric, the psychoactive mushroom *Amanita muscaria*.

BIRCH

Betula papyrifera

BETULACEAE (ALDER FAMILY)

Americans might not have invented birch beer, but we certainly perfected it. Birch trees are found all over North America, Europe, and Asia. Over the centuries they've been put to use as lumber and paper, as dye and resin, and as medicine. Archeologists have turned up drinking vessels in Europe dating back to 800 BC that contained birch sap residue, suggesting that it was used in wine making just as honey was.

Starting in the early seventeenth century, several scientists wrote about the use of birch sap in medicinal, or purely recreational, liquors. Flemish physician Johannes Baptista van Helmont wrote that birch sap could be collected in spring and poured "into the Ale, after the greatest settlement of its boyling or working, which Wines and Ales do voluntarily undergo in Hogs-heads." He recommended this naturally fermented sap as a treatment for ailments of the kidneys, urinary tract, and bowels.

A few decades later, in 1662, John Evelyn offered this recipe in *Sylva,* the first book on forestry ever published: "To every gallon of birch-water put a quart of honey, well stirr'd together; then boil it almost an hour with a few cloves, and a little limon-peel, keeping it well scumm'd: When it is sufficiently boil'd, and become cold, add to it three or four spoonfuls of good ale to make it work (which it will do like new ale) and when the yeast begins to settle, bottle it up as you do other winy liquors. It will in a competent time become a most brisk and spiritous drink."

But it was the American paper birch—the aptly named *papyrifera*—that yielded abundant, sweet sap just when early colonists could most use a drink. The settlers watched Native Americans tap birch trees in spring and capture the sap, but what they didn't see them do was make alcohol from it. In spite of abundant sources of sugar and grain, northern tribes did not seem to develop a tradition of brewing alcohol the way southwestern and Latin American native people did. But Europeans knew a good source of alcohol when they found it; they mixed the sweet sap and bark with water, honey, and whatever spices they could obtain to make a mildly alcoholic beer. Sassafras was often an ingredient; from this tradition the drink, sarsaparilla, became popular in Pennsylvania Dutch country.

By the time Prohibition drew near, brewers were creating nonalcoholic versions, called soft drinks, to get around the ban. Nonalcoholic birch beer remained a regional specialty throughout the twentieth century. Today the flavor returns in Root, the Pennsylvania-made liqueur based on the flavor of early American bark and root brews. A few wineries in the Scottish Highlands specialize in birch wine, and a Ukrainian vodka distiller employs the flavor in its Nemiroff Birch Special Vodka.

Birch sap can be used to produce xylitol, a natural sweetener that has been found to fight tooth decay, and the bark of some species are high in methyl salicylate, the primary constituent in oil of wintergreen. And as usual, the early physicians who prescribed the bark weren't entirely wrong: a birch extract called betulinic acid is being investigated as an anticancer drug.

CASCARILLA

Croton eluteria

EUPHORBIACEAE (SPURGE FAMILY)

This small, highly fragrant tree must have seemed like a natural addition to spirits. The bark's essential oil contains many of the same compounds found in pine, eucalyptus, citrus, rosemary, cloves, thyme, savory, and black pepper, making it attractive not just as a flavoring but as a bass note in perfumes as well.

The cascarilla tree is native to the West Indies and was described by Europeans in the late eighteenth century as part of a wave of botanical exploration. Any aromatic tree bark from the New World was being evaluated for its medicinal possibilities; this one was put to use in bitters and tonics of all kinds. It was originally described as a silver-barked tree, but botanists soon realized that the white color came from lichen that colonizes the tree. Under the lichen is a dark, corky bark that was used as a brown dye. Tiny sprays of pinkish white flowers and dark, glossy leaves make the tree an attractive specimen, but like other members of the spurge family (including poinsettia) the sap can be very irritating to handle.

Cascarilla bark continues to be an important ingredient in bitters and vermouth, and is rumored to flavor Campari as well. It has long been an additive to tobacco; when cigarette makers were required to disclose their ingredients in 1989, cascarilla was still on the list.

CINCHONA

Cinchona spp.

RUBIACEAE (MADDER FAMILY)

No tree has had a more important role in the history of cocktails than this South American species. The quinine extracted from cinchona bark doesn't just flavor tonics, bitters, aromatized wines, and other spirits. It also saved the world from malaria and put botanists and plant hunters at the center of several global wars.

Twenty-three different trees and shrubs make up the *Cinchona* genus, most of them sporting dark, glossy leaves and white or pink tubular, fragrant flowers visited by hummingbirds and butterflies. The reddish brown bark was used as a medicine by Andean tribes. They treated fevers and heart problems with it, and perhaps malaria—although some historians believe that malaria was introduced to South America by Europeans, who had suffered from it for centuries.

Jesuit priests discovered its efficacy against malaria in 1650, but it was a half century before Europeans grasped the importance of the bitter powder and started sending ships to South America to load up on felled trees. The locals were understandably concerned about the plunder of their forests and worked together to conceal the location of the trees.

Not every species of cinchona yields a potent dose of quinine, and botanical literature is full of misidentifications and misnamings of the trees. In 1854, a gorgeous book called *Quinologie* was published in Paris with hand-colored plates illustrating the different varieties so that pharmacists could tell the types of bark apart. We now know that *Cinchona pubescens* supplies the highest dose, as does *C. calisaya* and a few hybrids. While it may seem that *C. officinalis,* with its official-sounding name, would be the standard for quinine production, it actually contains very little of the drug.

THE MAMANI GIN & TONIC

Jalapeños and tomatoes, two South American natives,
pay tribute to Manuel Incra Mamani, the man who lost everything
to bring quinine to the rest of the world.

1½ ounces gin (try Aviation or Hendrick's)
1 jalapeño (or, if you prefer, a milder pepper), seeded, cored,
    and sliced
2 to 3 sprigs fresh cilantro or basil
1 cucumber (1 chunk and 1 swizzle-stick-shape slice needed)
High-quality tonic (look for a brand without high-fructose
    corn syrup, like Fever-Tree or Q Tonic)
3 red or orange cherry tomatoes

In a cocktail shaker, muddle the gin with 2 slices of the jalapeño,
1 sprig of cilantro, and the chunk of cucumber.

Fill a highball glass with ice; layer in 1 or 2 slices of jalapeño, a
sprig of cilantro, and the slice of cucumber.

Strain the gin and pour over ice. Fill the glass with tonic water;
garnish with cherry tomatoes on a pick.

Figuring this out was not easy for European explorers wandering the jungle, often in a fever themselves. One noted character in the quinine drama was a British merchant named Charles Ledger. In the 1860s, he sold a collection of seeds to the British government, but they turned out to yield little quinine. He hired a Bolivian man named Manuel Incra Mamani to gather more seed for him, but Mamani was captured by local officials. As Ledger himself described it: "Poor Manuel is dead also; he was put in prison by the Corregidor of Coroico, beaten so as to make him confess who the seed found on him was for; after being confined in prison for some twenty days, beaten and half starved, he was set at liberty, robbed of his donkeys, blankets and everything he had, dying very soon after."

Manuel did, however, manage to ship some seed to Ledger. By this point the British government would have no more of Ledger's schemes, so he instead sold them to the Dutch for the equivalent of twenty dollars. The Dutch sent them to Java, where they already had a long history of controlling spice plantations. Unlike the seeds Ledger sold to the British, these seeds proved viable, and soon the Dutch had a global monopoly. They developed an alternative to logging the trees: they would strip the bark and then wrap the trunk in moss to heal the wound so it could regenerate.

Everything changed in World War II, when Japanese troops took control of Java and Germans seized a quinine warehouse in Amsterdam. The last American plane to fly out of the Philippines before Japanese control carried four million quinine seeds—but the trees could not be grown fast enough to provide malaria remedies to Allied troops.

A desperate search for a synthetic alternative was under way, but meanwhile, American botanist Raymond Fosberg was sent by the USDA to South America to find more quinine. He traced the routes of the old explorers and managed to acquire 12.5 million pounds of bark to ship home—but it wasn't enough. One night in Columbia, Fosberg heard a knock at the door and discovered Nazi agents ready to make a deal. They'd followed him through South America and wanted to offer for sale a supply of pure quinine they'd smuggled from Germany. He didn't have to debate long before accepting their offer. American troops needed the drug if they were to keep fighting—even if it came from corrupt Nazis.

From its beginning as a medicine, one difficulty with quinine was its bitter taste. Mixing it with soda water, and perhaps a bit of sugar, was helpful. British colonists realized that adding a splash of gin improved the medicine considerably, and the gin and tonic was born. Quinine also became an important ingredient in bitters, herbal liqueurs, and vermouth. Byrrh (pronounced "beer") is a mixture of wine and quinine; Maurin Quina is a white wine aperitif infused with quinine, wild cherries, lemon, and cherry brandy. Italian aperitifs such as China Martini and Liquore Elixir di China are also quinine-based, as is the Spanish citrus liqueur Calisay. A wide range of quinine-infused aperitif wines are coming on the market or enjoying a renaissance; all of them are worth exploring.

Perhaps one of the loveliest uses of quinine in a drink can be found in a bottle of Lillet, a wine infused with citrus, herbs and a bit of quinine. Lillet, available in blanc, rosé, and rouge styles, is best enjoyed like wine, chilled in a glass, preferably while sitting in a French sidewalk café in spring—but bartenders are putting it to good use in cocktails as well.

WHY DOES QUININE GLOW UNDER ULTRAVIOLET LIGHT?

Shine a blacklight on a bottle of tonic water and it will glow a bright radioactive blue. The quinine alkaloid is "excited" by ultraviolet light, which means that the electrons absorb the light and take on extra energy, throwing them out of their regular orbit. In order to return to their natural position—their "relaxed" state—they release the energy, causing a bright glow.

CINNAMON

Cinnamomum verum
LAURACEAE (LAUREL FAMILY)

No one knows where cinnamon sticks come from. There is a bird called the cinnamon bird that gathers the fragrant twigs from some unknown location and builds its nest from them. To harvest the cinnamon, people attach weights to the tips of arrows and shoot the nests down.

That's not actually true, but it was Aristotle's best guess when he described cinnamon in his *Historia Animalium* in 350 BC. We have since located the source of cinnamon, relieving us of the necessity of shooting down the nests of mythical birds.

Cinnamon is, in fact, the bark of a tree native to what is now Sri Lanka. Arab spice traders had managed to keep the location a secret, but once Portuguese sailors found it, the news got out. They learned to wait until the rainy season to cut down young shoots, a practice called coppicing, which would stunt the growth of the tree and force it to put up one young trunk after another rather than grow into a fully mature tree. These pieces would be scraped to remove the gray outer bark, making it easier to cut long peels of the light inner bark. The peels would dry in the sun and roll into the curled pieces we buy today as cinnamon sticks.

Until the late 1700s, cinnamon was harvested from wild trees, but after that they were grown on plantations. Today the highest-quality cinnamon comes from Sri Lanka, but India and Brazil also supply the world market. It is generally labeled as true cinnamon or Ceylon cinnamon.

Another species of cinnamon tree native to India and China, *Cinnamomum aromaticum,* produces what is often called cassia cinnamon. It is widely sold in the United States and is easy to distinguish from true cinnamon: the sticks of cassia are thick and generally form a large double roll, whereas true cinnamon sticks look more like a tightly rolled bunch of thinner bark. The two are harder to tell apart after they've been ground into a spice powder, but there's a reason to check labels: cassia cinnamon can contain high levels of coumarin, which could cause liver damage to people who are sensitive to it. This makes Ceylon or true cinnamon a safer choice for anyone who has liver problems and plans to eat large quantities of the spice. Still, there is no ban or restriction on cassia as there is for tonka bean, another spice that contains a comparable amount of coumarin.

Cinnamon leaves are high in eugenol, also found in cloves. In the bark, the main component is called cinnamaldehyde, although the ever-present linalool, a spicy floral compound, is also there. Cinnamon is ubiquitous in the cocktail world: it's found in gin, vermouth, bitters, and spicy liqueurs. Perhaps the best-known cinnamon liqueur is Goldschläger, a clear cinnamon schnapps with bits of gold leaf floating in the bottle. The French distiller Paul Devoille makes a gingerbread liqueur called Liqueur de Pain d'Épices that is the perfect expression of cinnamon in a bottle.

DOUGLAS FIR

Pseudotsuga menziesii
PINACEAE (PINE FAMILY)

Inspired by the traditional pine liqueurs of Alsace, Portland distiller Stephen McCarthy wanted to make a spirit infused with his local conifer, the Douglas fir. This majestic evergreen grows to over two hundred feet along the Oregon coast, where it is the state tree. It is the host plant for a number of moths and butterflies, its sturdy wood is prized for timber, and it makes a fine Christmas tree.

To make his spirit, McCarthy went into the forest, handpicked buds from the tips of the branches, and tried, without success, to extract flavor from them. He was unable to create a drink to his liking, in part because the buds—young dark shoots that form the next year's needles—would oxidize as they were picked and handled.

Finally, he brought drums of his own high-proof neutral grape spirit into the woods, where he poured it into buckets and carried it directly to the trees. "We dropped the buds straight into the buckets," he said. "We were actually making the eau-de-vie out in the woods." He took the infused spirit back to the distillery, let it sit for two weeks, then filtered and redistilled the mixture. "Eau-de-vie is very unforgiving," he said. "It is not aged in oak, so any off flavors in the spirit or the ingredients will not be corrected in the barrel."

At last he was satisfied with the flavor of the finished product—but not the color. "It comes from an evergreen," he said. "It should be green. But the second distillation took all the color out." There was only one way to get the color he wanted: he took the second distillation back into the woods, poured it back into buckets, and carried those buckets back to the trees. "We picked buds again and dropped them into the buckets, and let them sit just long enough to bring the color back."

It took McCarthy years to figure out how to get the color, clarity, and flavor to come together, but his trials weren't over. He had to get federal approval for the label. "I wanted to put the Latin name of the Douglas fir (*Pseudotsuga menziesii*) on the label, because this product is all about the tree," he said. "But the federal liquor agents didn't believe that there was a tree called a Douglas fir, and they really didn't know what to make of the Latin name." Eventually the label, which also featured a drawing of the tree by his wife, the artist Lucinda Parker, won approval. McCarthy's Clear Creek Distillery now produces 250 cases per year of the green spirit.

THE DOUGLAS EXPEDITION

Stephen McCarthy prefers his Douglas fir eau-de-vie in a neat, single-ounce serving after dinner. But it makes a lovely cocktail as well. This one is named after David Douglas, the Scottish botanist who went on a famous 1824 plant-hunting expedition to the Pacific Northwest. He introduced almost 250 new species to England, including his namesake, the Douglas fir. Douglas died at the age of thirty-five while climbing a volcano in Hawaii. This drink pays tribute to his early days at London's Royal Horticultural Society, the group that sponsored his expeditions.

1 ounce **London dry gin**
1 ounce **Douglas fir eau-de-vie**
½ ounce **St-Germain elderflower cordial**
Juice of 1 **lemon wedge**

Shake all the ingredients with ice and serve in a cocktail glass.

EUCALYPTUS

Eucalyptus spp.
MYRTACEAE (MYRTLE FAMILY)

By 1868, the Tre Fontane Abbey near Rome was almost abandoned. The soil was depleted, the surrounding community deserted, and worst of all, malaria had reached an intolerable level. At that time people still believed that malaria was caused not by mosquitoes carrying a parasite but by something in the air; the word itself meant "bad air" in Latin. The monks hit upon an unusual solution to their problems: they planted stands of eucalyptus trees around the monastery. This fast-growing Australian tree, which, after all, smelled of medicine, would surely clear the air, rid the abbey of malaria, improve the soil, and give the monks some sort of crop with which to earn an income. They even made a tea from the leaves, which they believed would keep malaria away.

The American Medical Association ridiculed these efforts in an 1894 journal article titled "The Passing of the Eucalyptus." It pointed out that malarial outbreaks had taken place since the trees were planted and scoffed at its "reputed medical virtues." However, the monks were not entirely wrong: in 2011, an extract of *Eucalyptus citriodora* called oil of lemon eucalyptus won approval from the Centers for Disease Control and Prevention as a recommended mosquito repellant.

Still, the monks were left to contend with thousands of eucalyptus trees that served no real purpose. Like any good farmers, they found a way to bottle their crop. Today visitors to the abbey can pick up a bottle of sweet Eucalittino delle Tre Fontane, a liqueur made with macerated eucalyptus leaves. They also offer a bitter Estratto di Eucaliptus, made with no added sugar and recommended for cold winter nights.

Eucalyptus might seem like a flavor better suited to cough medicine than liquor, but the cool note of menthol or camphor can help amplify woodsy flavors like pine or juniper. It is used in bitters, vermouth, and gin. Fernet Branca in particular is known for its powerful eucalyptus flavor.

Eucalyptus has a long history as an intoxicating substance in its native Australia. The cider gum eucalyptus, *E. gunnii,* excretes a sweet, sticky sap that naturally ferments as it drips down the tree. As many as four gallons per day can flow from a single tree, and Aboriginal people made good use of it. In 1847, the British botanist John Lindley wrote that it "furnishes the inhabitants of Tasmannia with a copious supply of a cool, refreshing, slightly aperient liquid, which ferments and acquires the properties of beer." Today the Tamborine Mountain Distillery is winning awards for its Eucalyptus Gum Leaf Vodka and Australian Herbal Liqueur, both flavored with the leaves.

Bartenders are beginning to experiment with eucalyptus syrups and infusions, but it's important to note that only *E. globulus,* the so-called blue gum that is widely distributed in the western United States, is considered a safe food ingredient by the FDA—and it has only approved the use of the leaves, not the essential oil extract.

DRUNKEN LORIKEETS

EVERY YEAR, AUSTRALIAN ORNITHOLOGISTS FIELD CALLS about the strange behavior of the musk lorikeet population in the southeastern part of the country. These brilliantly colored parrots sometimes find themselves unable to fly. They stumble around on the ground and generally act like drunken louts. They even appear hung over the next day. It happens when their normal food source, eucalyptus nectar, ferments on the tree. This appears to be one of the only true accounts of wildlife being intoxicated by wild liquor. Unfortunately, it makes the birds vulnerable to predators or injury, so bird rescue organizations routinely take in drunken lorikeets and help them sober up.

MASTIC

Pistacia lentiscus

ANACARDIACEAE (CASHEW FAMILY)

A close relative to the pistachio, the mastic tree is native to the Mediterranean, where its resin has been harvested since ancient times for a surprising number of uses. Mastic gum oozes from the trunk when the bark is cut and dries to a hard, translucent yellow substance that softens to something like chewing gum when chewed. It is useful as a varnish; in fact, painters still use it on their canvases. As an adhesive, it's used in dissolvable stitches, bandages, and topical ointments. The gum seems to control cavities, so it's used in some toothpaste brands as well. And although the flavor is decidedly medicinal—imagine a cross between pine, bay laurel, and clove—it also flavors Greek spirits. Mastika is a high-proof anise-flavored spirit, usually with a brandy base, served as a digestif.

The tree itself is a small and highly fragrant scrubby plant with tiny red fruits that turn black with age. The Greek island of Chios is best known for its mastic gum production; in fact, mastic from Chios enjoys recognition from the European Union as a product with a protected designation of origin, much like Champagne or Calvados.

MAUBY

Colubrina elliptica
RHAMNACEAE (BUCKTHORN FAMILY)

Visitors to the Caribbean, particularly around Trinidad and Barbados, may have encountered mauby, a strangely sweet and bitter syrup made from the bark of two trees, *Colubrina arborescens* and *C. elliptica*. The recipe varies, but it usually involves mixing the tree bark with sugar, water, and some combination of cinnamon, allspice, nutmeg, vanilla, citrus peel, bay leaf, star anise, and fennel seed, giving it a spicy licorice kick. Poured into plain or soda water, mauby syrup has traditionally been considered a sort of cure-all. Islanders believe it treats diabetes and works as an appetite stimulant, but the only concrete proof of its health benefits comes from one small study in the *West Indian Medical Journal* that found that it could alleviate high blood pressure.

There are over thirty species of *Colubrina* around the world, all growing in warm-weather regions. *C. elliptica,* the species most widely used to make mauby, is actually native to Haiti and the Dominican Republic, but the bark circulates to nearby islands. *C. arborescens* is also used; it is native to Barbados. The wood is incredibly hard; in fact, the name ironwood has been applied to several species. The bark contains tannins and bitter saponins (called, in this case, mabioside) that probably protect the plant from predators. In Florida, *C. arborescens* is also called wild coffee, which suggests that the bark had an early history as a tea or coffee substitute.

In the early twentieth century, "mauby women" carried the homemade brew in tin containers perched on their heads and sold it informally on the streets. Now the syrup is manufactured on a larger scale and sold commercially, and bottled soft drinks like Mauby Fizz are also available. And mauby does, in fact, turn up in Caribbean cocktails, including those made by some of the best tiki bartenders in North America—although none of them will divulge their recipes.

MYRRH

Commiphora myrrha

BURSERACEAE (TORCHWOOD FAMILY)

M yrrh is an ugly little tree: scrawny, covered in thorns, and nearly bereft of leaves. It grows in the poor, shallow soils of Somalia and Ethiopia, where it is a gloomy gray figure in a barren landscape. If it weren't for the rich and fragrant resin that drips from the trunk, no one would give it a second look.

Tiny chunks of dried resin, about the size and shape of a raisin, were highly valued as perfume and incense among Egyptians, Greeks, and Romans. Because tree resin was used to seal drinking vessels, it is easy to see how myrrh and wine came together. Romans offered a myrrh and wine blend during crucifixions; it was presented to Jesus, but he refused it.

Myrrh has a bitter and somewhat medicinal taste. Its essential oil contains compounds also found in pine, eucalyptus, cinnamon, citrus, and cumin. The French distiller Combier lists it as an ingredient in its premium orange liqueur Royal Combier, and it is a common ingredient in vermouth, aromatized wines, and bitters. The makers of Fernet Branca do not bother to hide the fact that it is among their secret ingredients; the powerful and ancient flavor of myrrh helps explain why Fernet packs such a punch.

❧ PINE ❧

Pinus spp.
PINACEAE (PINE FAMILY)

The residue of wine laced with tree resin has been found at archeological sites dating back to the Neolithic period. Perhaps it was used as a preservative or to add a woodsy flavor the way barrel-aging would. It might have also had medicinal uses: tree resin seemed to heal trees, so it stood to reason that drinking tree resin might heal internal ailments. Roman winemakers added a potpourri of strange ingredients to wine, including not only pine resin but also frankincense, myrrh, and an extract from the terebinth tree, from which turpentine was also made.

Even today, it is possible to find a pine-resin-infused wine called retsina in Greece. The Greek winery Gaia Estate makes a retsina called Ritinitis Nobilis that is flavored with an extract of the Aleppo pine *Pinus halepensis.* Fernet Branca is also rumored to contain a touch of pine resin.

But the most interesting pine-based spirit is surely the Alsatian pine liqueur called *bourgeon de sapin.* It might not be to everyone's liking, but it is a rare and unusual historic liqueur that bartenders love to experiment with. (Imagine a boozy, sugary Christmas tree, enjoyed neat in a short glass.) An Austrian version called Zirbenz Stone Pine Liqueur gets its pale cinnamon hue and floral perfume from the arolla stone pine, *Pinus cembra,* which grows high in the Alps. According to the distiller, the cones are only harvested every five to seven years, and even then, less than a quarter of the cones are picked. The work is done by intrepid mountaineers who trek through the Alps in early July and climb the dense trees to reach the cones precisely when they are at their reddest and most pungent.

ROYAL TANNENBAUM

(Based on Lara Creasy's recipe, *Imbibe* magazine, Nov./Dec. 2008)

1½ ounces London dry gin
½ ounce pine liqueur, such as Zirbenz Stone Pine Liqueur
1 sprig fresh rosemary

Shake the gin and pine liqueur over ice, and strain into a cocktail glass.
Garnish with the rosemary sprig.

SENEGAL GUM TREE

Senegalia senegal (syn. *Acacia senegal*)

FABACEAE (BEAN FAMILY)

A small, prickly tree that grows in the Sudanese desert has been responsible for such diverse duties as keeping newspaper ink on the page, preserving Egyptian mummies, and stabilizing the sugar and color in soft drinks. It is also the key ingredient in old-fashioned *gomme* syrup, where it adds a smooth, silky texture to cocktails and keeps the sugar from crystallizing.

There were, until recently, over a thousand species of trees classified as acacias, almost all of them coming from Australia. A few are native to warmer areas of Europe, Asia, Africa, and North and South America. But taxonomists recently split the acacias into several different genera, a decision that was so controversial that petitions were circulated, botanists were sniping at one another publicly, and accusations of greed and corruption were flung at scientists during the normally staid conference on botanical nomenclature. As a result of their reorganization of the acacia genus, Sudanese farmers no longer grow acacias; they now grow *Senegalia senegal*. Acacia gum, also called gum arabic, will presumably have to find another name as well.

The botanical debate is not the only controversy surrounding the tree. Because it grows in Sudan, it is at the center of a brutal war. Supplies of the raw gum—which is harvested by scraping the tree and hand-collecting the globs of gum that issue forth—have been threatened by fighting in the region. The U.S. State Department issued warnings in 1997 about the possibility that Osama bin Laden had invested heavily in the Gum Arabic Company, a government-controlled monopoly that exported the gum to Europe for processing. The company denied ties to the terrorist. After heavy lobbying from the soft drink industry, the economic sanctions imposed on Sudan were amended to grant an exemption for gum arabic.

Another threat to the tree is climate change: as drought conditions worsen, it is confined to an ever-smaller belt across Sudan. Agricultural aid workers are trying to expand the tree's habitat and teach farmers how to grow "gum gardens" using special water catchment techniques that help the trees survive on minimal rainfall and produce enough gum to support a family. The farmers also contend with plagues of locusts, termites, fungal diseases, and ravenous goats and camels.

This twenty-foot-tall tree puts down a taproot one hundred feet long, which explains its ability to survive tough desert conditions. The tiny leaves help prevent water loss, and the canopy's expansive umbrella shape gives those leaves maximum exposure to sunlight to make up for their size. The sweet, sticky gum also serves a purpose to the tree, healing wounds, protecting against insect damage, and fighting disease.

By about 2000 BC, Egyptians had learned that scraping the tree's bark would put it under stress and force it to produce more gum. They used it to make ink, mixed it into food, and employed it as a binding paste for mummification. (The old French word *gomme* comes from older words for gum, the Egyptian *komi* and Greek *komme*.) The gum has been in continuous use as a binder in inks,

GOMME SYRUP

2 ounces **powdered food-grade gum arabic**
6 ounces **water**
8 ounces **sugar (less to taste)**

Combine the gum arabic and 2 ounces of the water in a saucepan and heat to a near boil, dissolving the gum. Once cooled, make the simple syrup by combining the sugar and remaining 4 ounces of water in a saucepan. Bring to a boil, allowing sugar to dissolve. Add the gum mixture, heat for 2 minutes, and then allow to cool. Some people prefer a simple syrup made with equal parts sugar and water, so try a small batch and then adjust the quantities to your taste. Keep refrigerated; it should last at least a few weeks.

paints, and other products of industry and as a thickener and emulsifier for medicinal syrups, pastes, and lozenges. Bakers used it in ice cream, candies, and icing, and it was only a matter of time before the sweet *sirop de gomme* became a useful cocktail ingredient as well. It adds a silky texture that is impossible to replicate with simple syrup.

Gum arabic has returned as an ingredient in specialty cocktail syrups, but it's also easy to mix up a batch at home. Buy food-grade gum arabic from spice stores or shops catering to bakers and confectioners. (Gum arabic sold in craft stores is a lesser grade intended for use in art projects.)

SPRUCE

Picea spp.
PINACEAE (PINE FAMILY)

T he fact that a vitamin C deficiency caused scurvy was not entirely understood until the 1930s, but ship captains had, from time to time, managed to prevent the disease by stocking up on lemons and limes before leaving on a long voyage. And when citrus wasn't available, they unknowingly substituted other sources of vitamin C—including the young, green tips of spruce trees.

Captain James Cook tried a recipe on his crew that he had obtained from the botanist Joseph Banks. The recipe consisted of spruce twigs boiled in water with some tea to improve the flavor, which would then be combined with molasses and a bit of beer or yeast to start the fermentation. Cook wrote in his journal that either berries or spruce beer cured the crew of scurvy.

Spruce beer was well known to Jane Austen, who wrote to her sister Cassandra in 1809 about brewing a "great cask" of it. A pivotal moment in *Emma* even revolves around a recipe for spruce beer, with Mr. Knightley offering the recipe to Mr. Elton, who borrows a pencil from Emma to write it down. In a classic Austen plot twist that would prove important later, her friend Harriet stole the pencil Mr. Elton used to write down the ingredients as a remembrance of him.

Recipes for spruce beer were abundant in eighteenth- and nineteenth-century journals. Benjamin Franklin is widely credited with creating a recipe for the beer—but it wasn't his invention. While he was ambassador to France, he copied several recipes from a cookbook called *The Art of Cookery Made Plain and Easy,* written by a woman named Hannah Glasse in 1747. (Glasse, by the way, wrote a number of interesting recipes Franklin missed, including one called Hysterical Water that included parsnips, peony, mistletoe, myrrh, and dried millipedes, soaked in brandy and "sweetened to your taste.") He never meant to take credit for her recipe; he simply copied it for his personal use. Nonetheless, it was found among his papers, and the story that one of the Founding Fathers created a recipe for spruce beer was too good to resist. Modern re-creations of the recipe credit him alone, not Hannah Glasse.

Spruce trees are ancient creatures, dating to the late Jurassic period, 150 million years ago. There are as many as thirty-nine species, depending on which botanist you ask, and they are distributed across colder climates in Asia, Europe, and North America. Like many conifers, the trees grow slowly and, if left unmolested by chainsaws, live to an astonishing old age. The world's oldest living tree is a Norway spruce with a root system that dates back 9,950 years.

The trees produce ascorbic acid, and other nutrients that help fight scurvy and enhance uptake of vitamin C, as a defense mechanism in winter to help them survive the cold and develop seed cones. Highest levels of the vitamin are found in the red and black spruce (*P. rubens* and *P. mariana*), but the FDA has approved only black spruce and the white spruce, *P. glauca,* as safe natural food additives. To the untrained eye, spruce trees can resemble other, highly poisonous conifers such as the yew, so home brewers are advised to get expert advice before harvesting in the forest.

SUGAR MAPLE

Acer saccharum

ACERACEAE (MAPLE FAMILY)

In 1790, Thomas Jefferson bought fifty pounds of maple sugar to sweeten his coffee. This was less a culinary decision than a political one: he'd been pressured by his friend and fellow signer of the Declaration of Independence, Dr. Benjamin Rush, to advocate for the use of home-grown maple sugar instead of cane sugar, which was dependent upon slave labor.

Although he was a slave owner himself, Jefferson nonetheless saw the wisdom behind this idea. He wrote to a friend, British diplomat Benjamin Vaughan, that large swaths of the United States were "covered with the sugar maple, as heavily as can be conceived," and that the harvesting of maple sugar required "no other labor than what the women and girls can bestow . . . What a blessing to substitute a sugar which requires only the labor of children, for that which is said to render the slavery of the blacks necessary."

The possibility of replacing slave labor with child labor was not the only reason early Americans were excited about maple sugar. It was seen as a rich and healthy sweetener—and in fact, maple syrup contains iron, manganese, zinc and calcium, along with antioxidants and a wide range of volatile organic flavors, giving it notes of butter, vanilla, and a warm, woodsy spiciness that is also found in spirits aged in oak. Although Dr. Rush, who also advocated temperance, would not have approved of it, maple syrup made a fine brew. A few people reported that they saw Iroquois people making a lightly fermented drink from the sap, although that would have been unusual for a northern tribe, where alcohol was uncommon before European contact. But settlers certainly got down to business: an 1838 recipe involved boiling down maple sap, mixing it with wheat

bar

PART II

257

TREES

or rye when barley was not available, adding hops, and aging it in casks after fermentation.

The sugar maple *Acer saccharum* is native to North America and one of about 120 species known around the world. Most maples are actually native to Asia (such as *A. palmatum,* the popular red-leaved Japanese maple), and while there are many European species, none yielded such a remarkably sweet sap. It wasn't until settlers saw Iroquois tapping maple trees for sugar that they saw the potential.

The sugar maple tree is unique in that its sapwood—the outer part of the trunk that is still growing—contains hollow cells that fill with carbon dioxide during the day. On a cold night, the carbon dioxide shrinks, creating a vacuum that pulls sap up the tree. If the weather is warm the next day, the sap flows down again—and this is when maple farmers know to tap the tree. The sap is boiled to make syrup, and it can be further heated to turn it into granulated sugar.

Quebec is known for its maple tradition. A popular winter drink called the Caribou is made from wine, whiskey, and maple syrup. Maple-infused whiskey liqueurs and eaux-de-vie from the region are well worth sampling, as are maple wines and beers. Vermont turns out good maple spirits, too, including a fine maple vodka from the endlessly inventive Vermont Spirits, a distillery that has also helped keep Vermont dairies in business with its Vermont White, a vodka distilled from milk sugars.

CARIBOU

3 ounces **red wine**
1½ ounces **whiskey or rye**
Dash of maple syrup

Shake all the ingredients over ice and strain. One variation on this recipe involves equal parts port and sherry, a splash of brandy, and a dash of maple syrup. Experiment at will, but please use real maple syrup, not the imitation.

-- proceeding onward to --
FRUIT

fruit:

THE RIPENED OVARY OF A FLOWER,

FORMED AFTER OVULATION,

GENERALLY CONSISTING OF

A FLESHY OR HARD OUTER WALL

SURROUNDING ONE OR MORE SEEDS.

⋟ APRICOT ⋞

Prunus armeniaca
ROSACEAE (ROSE FAMILY)

P our yourself a glass of amaretto and you'll recognize the flavor immediately: almonds. Right? Not necessarily. The world's most popular amaretto, Amaretto di Saronno, gets its almond flavor from the pits of apricots.

Just as almonds may be sweet or bitter—the bitter varieties containing high levels of amygdalin, which turns into cyanide in the gut—apricot pits are also classified as sweet or bitter. Most varieties grown in the United States are cultivated for their fruit, and their pits are the bitter variety. But in the Mediterranean, it is easier to find so-called sweet pit or sweet kernel varieties. Split open the hard pit of a sweet variety and the kernel inside—the seed— looks and tastes much like the closely related sweet almond.

The apricot was cultivated in China around 4000 BC; by 400 BC, farmers were selecting specific varieties. It arrived in Europe over two thousand years ago. There are hundreds of varieties now, many of them uniquely adapted to a specific region. One of the oldest sweet kernel cultivars is Moor Park; it dates to at least 1760 in England. The most popular cultivar before Moor Park was called Roman. It was actually developed in ancient Rome.

The tradition of flavoring alcohol with apricots seems to have begun about ten minutes after the introduction of the apricots themselves. Some of the earliest recipes for ratafia call for soaking apricot kernels in brandy with mace, cinnamon, and sugar. The invention of amaretto was not far behind; many of them are still made with

VALENCIA

In 1927, the International Bartenders Union gathered in Vienna for a cocktail competition. The winner was a German bartender named Johnnie Hansen, whose drink was a mixture of apricot brandy, orange juice, and orange bitters. The European bartenders sent the news to the United States with a tip of the hat to the Anti-Saloon League, thanking it for its work to advance the cause of Prohibition, which would only bring more drinkers to Europe.

The Valencia made it into the classic 1930 *Savoy Cocktail Book*. Here it is with an Austrian liqueur made from actual apricots. Freshly squeezed juice is, of course, essential.

1½ ounces Rothman & Winter Orchard Apricot Liqueur
¾ ounce freshly squeezed orange juice
4 dashes orange bitters
Orange peel

Shake all the ingredients except the orange peel over ice and strain into a cocktail glass. Garnish with the orange peel. The Savoy Cocktail Book *calls for pouring it into a highball glass and topping it with a dry sparkling* cava *or Champagne. Even better, perhaps, is a variation suggested by Erik Ellestad, a cocktail writer whose Savoy Stomp blog (savoystomp.com) documented his journey through the entire* Savoy Cocktail Book. *He recommends equal parts (¾ ounce each) orange juice, apricot liqueur, and Armagnac, with Angostura instead of orange bitters, then topped with* cava.

apricot kernels rather than almonds. In France, the word *noyau* (or the plural, *noyaux*) refers to the pit of a stone fruit, and in practice a liqueur of that name is typically made with apricot pits. Crème de noyaux turns up in old cocktail recipes but is nearly impossible to find in the United States. The French distiller Noyau de Poissy makes two versions, but a trip to France may be required to secure a bottle.

The fruit itself is, of course, also used to make brandy, eau-de-vie, and liqueurs; in Switzerland, apricot spirits are called *abricotine.* Under the modern usage of the word *brandy,* a spirit called apricot brandy would be distilled from the apricots themselves. But in the nineteenth and early twentieth centuries, apricot brandies and peach brandies were made from grape-based brandy with fruit juice added. In fact, a 1910 case involving adulterated apricot brandy made with imitation ingredients was one of the early enforcement actions taken under the U.S. Pure Food and Drug Act of 1906. This historical detail is of some importance to cocktail enthusiasts attempting to re-create Prohibition-era drinks: a recipe calling for apricot brandy (or peach brandy, for that matter) might actually refer to something more like a sweet liqueur than a dry, higher-proof brandy.

BLACK CURRANT

Ribes nigrum

GROSSULARIACEAE (SYN. SAXIFRAGACEAE)
(GOOSEBERRY FAMILY)

Writing in the twelfth century, St. Hildegard recommended the leaves of the cassis plant as a cure for arthritis. "If one suffers from the gout," the abbess, botanist, and philosopher wrote, "take in equal part cassis leaves and comfrey, crush them in a mortar, and add the grease of wolf." While mixing the plant with wolf grease was one way to cure one's ills, mixing it with alcohol proved far more popular. Black currants—called cassis in France—are the sole flavoring in the dark red, syrupy sweet liqueur known as crème de cassis.

The European black currant is not native to Dijon, France—it comes from colder northern European countries and parts of northern and central Asia—but farmers in Dijon perfected the art of coaxing the plants into producing smaller fruit with a deeper, richer color and a more intense flavor.

Apart from medieval remedies, the earliest liqueur made from the fruit was ratafia de cassis, a mixture of brandy and black currants that was allowed to soak for six weeks, and then was strained and blended with a sugar syrup. Today, crème de cassis is made by crushing the fruit and macerating it in straight alcohol—usually a neutral grape spirit—for two months. Then the fruit is pressed to expel the remaining juice and strained. The liqueur is piped into another vat, then mixed with beet sugar and water to adjust the sweetness and get the alcohol to about 20 percent ABV.

A one-quart bottle might contain the extract of just under a pound of fruit. In the case of higher-end "supercassis" liqueurs, the amount of fruit is doubled or tripled to make a thicker and fruitier

CURRANT or RAISIN?

In the United States, the word *currant* is sometimes used to refer to a small, seedless raisin. Those are dried grapes unrelated to currants in the *Ribes* genus.

drink. To judge the quality of a bottle of crème de cassis, shake it and observe how the liqueur coats the glass. A supercassis will leave behind a thick, burgundy-colored syrup. Cooks in Dijon don't just drink it; they also pour it over fromage blanc and use it in *bœuf bourguignon*.

Crème de cassis enjoyed a boost in popularity in the late nineteenth century. It was common in French cafés to simply place a bottle on every table and let patrons add it to their own drinks. After World War II the mayor of Dijon, Félix Kir, poured a drink for visiting dignitaries that consisted of crème de cassis and white wine. The drink, which became known worldwide, is now called a Kir in his honor.

The true medicinal uses of black currants became better known around that time as well. With oranges in short supply in Britain during and after World War II, a black currant juice called Ribena was distributed free to children. High in vitamin C, antioxidants, and other healthy compounds, it kept many children from malnutrition. The fruit is still marketed as a "superfood" that may have a number of disease-fighting benefits.

Black currants, and the liqueurs made from them, are not well known in the United States, owing in part to a strange quirk in the agricultural laws. The plant acts as host to a disease called white pine blister rust that kills eastern pine trees. The disease cannot spread from one pine tree to another; it must first make a stop on a currant bush, where it produces a particular kind of spore that allows it to reinfect a pine tree. In the 1920s, the timber industry lobbied to have the currant banned, in spite of the fact that simple forest management practices could have interrupted the disease cycle. The spores can travel up to 350 miles from pine tree to currant bush but can only travel a thousand feet from currant bush back to pine tree. This makes it fairly simple to stop the disease from spreading; foresters simply have to keep the currants at least one thousand feet away from the trees. It also helps that at least 20 percent of pines are naturally resistant, and the rest will only become infected if the weather is particularly damp when the spores are traveling.

The ban was lifted nationwide in 1966, but many states kept the restriction in place. Cornell agriculturalist Steven McKay, who had fond memories of currants from his travels in Europe as a student, has worked to abolish the restriction and encourage farmers to

WHY DOESN'T CRÈME DE CASSIS CONTAIN CREAM?

CRÈME DE: In Europe, the term *crème de* followed by the name of a fruit refers to a liqueur with a minimum sugar content of 250 grams of invert sugar (a kind of sugar syrup) per liter and an alcohol content of at least 15 percent ABV. Crème de cassis, however, must contain at least 400 grams of invert sugar per liter.

CRÈME: In the past, some very sweet liqueurs were sold as "crème cassis" or "crème (name of fruit here)" to indicate an even higher sugar content. There is no official legal definition for this term, but it is generally meant to convey an especially sweet liqueur.

CREAM: A liqueur with the word *cream* on the bottle, such as Irish cream, contains milk solids.

LIQUEUR: In the United States, the term *crème* is, according to legal definitions, replaced by the term *liqueur* or *cordial*, which refers to any flavored, sweet distilled spirit containing at least 2.5 percent sugar by weight.

BLACK CURRANTS

The European black currant is a stiff, upright shrub that reaches about six feet tall and produces clusters of fruit resembling tiny bunches of grapes. They do best in rich, moist, slightly acidic soil with regular mulching, prefer full sun and regular water, and are hardy to −25 degrees Fahrenheit.

The fruit appears only on year-old canes, which means that new growth should be left alone for a full year so that it can bear fruit. Pick currants while they are dry and firm; a mature bush may produce ten pounds per year. In winter, cut two to four older canes to the ground, and choose a few older branches to trim back to the point where younger side shoots emerge. If the bush stops fruiting, cut it all the way to the ground and wait two years for fruit.

Check with a local fruit nursery to choose a variety best suited to your climate and most resistant to local diseases or pests. Noir de Bourgogne is the variety most often used for the French liqueur, but it is hard to find in the United States and not well suited for every climate. Ben Lomond and Hilltop Baldwin are two good, vigorous cultivars. Native American black currants, including the clove currant *Ribes odoratum* and the American black currant *R. americanum*, both produce edible berries but have not been used as much in liqueurs.

Red and white currants are also worth growing, if only as a cocktail garnish or a snack to eat right off the bush. The pearly Blanca white currant can be used to make a currant wine, and Jonkheer Van Tets is considered one of the most vigorous and flavorful red currants.

FULL/PART SUN

LOW/REGULAR WATER

HARDY TO -25F/-32C

grow the crop. Now disease-resistant varieties, modern fungicides, and a better understanding of how the disease is transmitted make blister rust a thing of the past. Still, several states on the East Coast continue to outlaw the growing of currant bushes.

In Europe, black currants were involved in another notable legal tangle. Cassis was central to one of the most important court cases in the early days of the formation of the European Union. In France, crème de cassis was bottled at 15 to 20 percent ABV, but an exporter found that it could not be sold as "liqueur" in Germany because in that country, the alcohol content had to be at least 25 percent. The court case that followed in 1978, now referred to as the Cassis de Dijon case, determined that laws established in one member country must be recognized in another, establishing the principle of mutual recognition that paved the way for more robust trade among EU nations.

KIR

4 ounces dry white Burgundy such as Aligoté,
    or another dry white wine
1 ounce crème de cassis

Pour the cassis in a wine glass, and add the white wine. Adjust the portions to taste. A Kir Royal uses Champagne instead of wine; a Kir Communiste, made with a red wine, calls for beaujolais; and a Kir Normand mixes the liqueur with cider. For a lighter drink, mix 1 part crème de cassis with 4 parts sparkling water.

CACAO

Theobroma cacao

MALVACEAE (MALLOW FAMILY)

The cacao is the most unlikely of fruits. It comes from a tropical evergreen tree that prefers to grow within 10 degrees' north and south latitude of the equator. When mature, it produces ten thousand blossoms in a season. Fewer than fifty of those flowers will ripen into fruit, and only if they have been pollinated by flying midges or particular species of ants.

The fruit takes the form of an enormous pod the size and shape of a football. Each pod contains up to sixty beans surrounded by a soft pulp. The pulp is tempting to birds and monkeys because it is so rich in sugar and fat. The beans themselves are not as interesting to mammals because of their bitter flavor, so they get left behind to go to seed.

Jungle animals aren't the only ones who like the juicy pods. If they are left alone on the ground, cacao will ferment spontaneously. Spanish explorers were surprised to arrive in Guatemala and see dugout canoes filled with cacao fruit. The fruit would ferment until the bottoms of the canoes were filled with "an abundant liquor of the smoothest taste, between sour and sweet, which is of the most refreshing coolness." The Spaniards came looking for gold, but they found chocolate, the next best thing.

It is no small miracle that chocolate and booze arise spontaneously in nature. Even today, chocolate is made by fermenting the beans for several days to allow richer and more complex flavors to emerge. They are then dried, roasted, and cracked open so that the nibs—the meaty part of the bean—can be extracted. The nibs are ground into a powder or paste that, along with a little sugar, becomes dark chocolate. If milk is added, it becomes milk chocolate. And if the fat, called cacao butter, is extracted by itself and mixed with sugar, that is white chocolate.

Today chocolate can be found in any number of syrupy sweet liqueurs. Unfortunately, far too many bars sell a dreadful concoction known as a chocolate martini. Drink these if you must, but there are far more subtle and sophisticated ways to enjoy chocolate spirits. Dogfish Head makes a cacao beer called Theobroma that is intended to be a modern recreation of an ancient Olmec recipe. Based on residue analysis of pottery dating to 1400 BC, plus some hints from the reports of Spanish explorers, their recipe includes honey; chili pepper; vanilla; and annatto, a reddish spice derived from the achiote tree, *Bixa orellana,* which is also used as a natural food coloring for cheese and other processed foods. The beer is earthy and spicy with just a hint of chocolate.

A more elegant and modern use of cacao in spirits comes from New Deal Distillery in Portland, whose Mud Puddle is an unsweetened infusion of roasted cacao nibs in vodka, resulting in a pure chocolate flavor without a trace of cloying sweetness.

FIG

Ficus carica

MORACEAE (MULBERRY FAMILY)

The fig tree is a strange and ancient creature. What most of us would call the fruit of a fig is not a fruit at all, but a syconium—a teardrop-shaped bit of plant flesh that contains inside it clusters of tiny flowers. The only way that we could see the flowers would be to split it open, but diminutive fig wasps know how to crawl in through tiny openings and pollinate the flowers. The fruit produced by these flowers are actually the fleshy, stringy tissue that we see when we bite into that thing we call a fig.

Confused? That's not all. Some figs must be pollinated by a wasp in order to set seed and reproduce, but the wasp lays her eggs inside that fruitlike structure and often dies inside. That means that the fig contains bits of wasp corpses—which is not very appealing. But around 11,000 BC, someone noticed that some fig trees could bear fruit without any pollination at all. Of course, the lack of pollination meant they couldn't reproduce, so people had to take cuttings to help them survive—and they did, for thousands of years.

Thanks to the efforts of our Middle Eastern Stone Age ancestors, we don't have to eat figs filled with the body parts of wasps, nor do we have to pick them out of our distillation equipment. Today's figs are either not pollinated at all, or they produce longer flowers that allow the wasps to do their work without actually crawling inside.

Figs came to Mexico in 1560 and have been planted in warmer climates all over the world, with hundreds of varieties in cultivation. Dried figs have always been useful as a source of portable, long-lasting nutrition: they contain a respectable amount of protein as well as essential vitamins and minerals.

Like almost any fruit, figs are distilled. A fig brandy called *boukha* comes from Tunisia, and in Turkey, a clear anise-flavored spirit called raki can be made from figs. A 1737 recipe for fig liqueur involved steeping figs in brandy along with nutmeg, cinnamon, mace, saffron, and licorice, "'till the whole virtue be extracted from them." An even stranger recipe from that era called for boiling snails with milk, brandy, figs, and spices, and offering it to people with consumption. Even if it didn't treat their illness, it would certainly give them something else to worry about.

Fortunately, modern fig liqueurs are much improved: look for French *crème de figue,* fig arak, and black fig-infused vodka, as well as local eaux-de-vie made anywhere figs are grown.

MARASCA CHERRY

Prunus cerasus var. *marasca*
ROSACEAE (ROSE FAMILY)

In the distant, boozy past, a maraschino cherry was not an artificially dyed and overly sweetened atrocity. It was a dense, dark, sour cherry called the marasca that grew particularly well in Croatia, around the town of Zadar. That region was known for the practice of fermenting marasca cherries with a little added sugar to produce a clear spirit called maraschino liqueur. Cherries could then be soaked in that liqueur to preserve them—and that is a proper maraschino cherry.

To understand why we associate maraschino cherries with Italy requires a brief history lesson. Given Zadar's advantageous location as a port city on the Adriatic Sea, it found itself under near-constant attack and was under the control of almost every nearby country at one time or another. The Luxardo company, the best-known maker of maraschino liqueur, has a history that mirrors that of the region: founded in Zadar in 1821, the distillery was at the center of nonstop political upheaval until World War I, when Italy took control. Many Croatian farmers, finding themselves Italian citizens, did the only sensible thing and decamped for Italy, taking cuttings from their cherry trees—and their recipes—with them.

After repeated bombings during World War II, the Luxardo distillery was decimated. Only one Luxardo family member survived; he, too, went to Italy to rebuild the business. Now many Italian distilleries make a version of maraschino liqueur, owing in part to Croatia's war-torn history.

In 1912, an early version of the FDA called the Board of Food and Drug Inspection issued a ruling that only marasca cherries preserved in maraschino could be labeled "Maraschino cherries." American growers favored large sweet cherries (a different species, *Prunus avium*), and they had developed a brining process that involved bleaching them in sulfur dioxide, which removed all the color but could also turn them to mush. To solve that problem, they added calcium carbonate (widely available at plaster and paint stores in those days) to harden them. What was left was described in one American agricultural report as nothing but bleached cellulose "in the shape of a cherry" that was then dyed red with coal tar, flavored with a chemical extract of stone fruit called benzaldehyde, and packed in sugar syrup. This product, whatever it was, could not be called a maraschino cherry.

But that changed, thanks in part to Prohibition. The temperance movement, working with soda manufacturers, campaigned against the evils of European cherries soaked in liquor. They championed the chemically treated, alcohol-free "American cherry, without foreign savor and without entangling alliances" over a "distillate of some foreign province, from fruit gathered by underpaid peasants, and handled and sold under conditions which would disgust the purveyors and purchasers of such products." Thanks to their efforts, real marasca cherries in pure liquor become disgusting in the minds

HOMEMADE MARASCHINO CHERRIES

Clean and pit fresh cherries (sour, if possible).

Loosely fill a clean Mason jar with the cherries.

Pour maraschino liqueur (or brandy or bourbon) over the cherries until they are completely covered.

Seal the jar, refrigerate, and use within 4 weeks.

of Americans, and bleached and dyed cherries became wholesome. In 1940, the FDA gave up the fight and agreed that any chemically treated, artificially dyed batch of cherry-shaped cellulose in a jar could be sold as maraschino cherries. (To add insult to injury, the FDA allows up to 5 percent of the cherries in a jar to contain maggots, calling that an "unavoidable defect.") Fortunately, authentic maraschino cherries, made by Luxardo and other companies, are available from specialty food shops today as an alternative—and they are easy to make at home.

Sweet cherries are native to either Asia or central Europe; early archeological evidence points to both locations. By Roman times, at least ten varieties were in cultivation. Sour cherries have also been cultivated in Europe for at least two thousand years.

Although cherries have been grown across the United States, they found their most advantageous climate in Oregon. One early pioneer in the Oregon cherry business was Seth Lewelling, who came from Indiana with his family in 1850. Lewelling was an abolitionist who helped organize a local chapter of a new anti-slavery political

A FIELD GUIDE
TO CHERRY-BASED SPIRITS

As with almost any fruit, cherries can be fermented and distilled in endless variations. Here are just a few worth trying:

CHERRY BRANDY usually describes a cherry liqueur, meaning a maceration of cherries and sugar in a base spirit such as brandy. Cherry Heering is a fine example; it's flavored with almonds and spices. American Fruits Sour Cherry Cordial is another outstanding cherry liqueur.

CHERRY WINE is a wine made from cherries rather than grapes. Maraska cherry wine from Croatia is the best known, and perhaps most authentic, version.

GUIGNOLET is a French cherry liqueur usually made from the large, sweet red or black *guigne* variety.

KIRSCH or **KIRSCHWASSER** is a clear brandy or eau-de-vie fermented with cherry pits, which impart a mild almond flavor. Made in Germany, Switzerland, and elsewhere; sometimes simply sold as cherry eau-de-vie.

MARASCHINO is a not particularly sweet liqueur made of a distillate or maceration of marasca cherries, usually double-distilled to make it clear. Luxardo is one of several distilleries making maraschino liqueur.

party called the Republican party. For his opposition to slavery, he was labeled a "black Republican." He told his critics that he would make them relish that term, and he did it by naming a new variety of cherry Black Republican so that they would have no choice but to eat their words. The Black Republican was once the most popular cherry for canning and preserving, but now Royal Ann and Rainier are more common.

THE (HYBRIDIZED) BROOKLYN COCKTAIL

1½ ounces rye or bourbon
½ ounce dry vermouth
¼ ounce Maraschino liqueur
2 to 3 dashes Angostura or orange bitters
1 maraschino cherry

Stir all the ingredients except the cherry with ice, strain into a cocktail glass, and garnish with the cherry. Purists will object that a Brooklyn is traditionally made with Amer Picon, a bitter orange aperitif, not with Angostura or orange bitters. If you have access to some Amer Picon, by all means, add ¼ ounce. Otherwise, this variation is quite nice and makes good use of marasca cherries in two forms.

CHERRY TREE

FULL SUN

LOW/REGULAR WATER

HARDY TO -25F/-32C

There are at least 120 species of cherry tree, many of which are not grown for their fruit. The flowering cherry trees on display in Washington, DC, in the springtime, for instance, are mostly *Prunus x yedoensis* 'Yoshino Cherry' and *P. serrulata* 'Kwanzan', two Japanese species. Most cultivars produce small, inedible fruit or are sterile so produce no fruit at all. The sour cherry, *P. cerasus*, is incapable of interbreeding with sweet cherries and is, in fact, self-fertile, which means that it doesn't need another tree nearby for pollination.

Sour cherry varieties are broadly divided into morellos, which are darker, and amarelles, which are lighter. There are hundreds of cultivars of each, most adapted to a particular climate. The marasca is a type of morello not widely sold in the United States, but backyard orchardists can easily substitute another sour cherry, such as Montmorency, North Star, or English Morello, that does well in their region.

Cherry trees are sold on dwarf or full-sized rootstock. It's important to choose the rootstock for the space available. Remember that birds love to pick ripe cherries off a tree, so a dwarf tree might be easier to protect with netting. Be sure to find out if another tree is needed for pollination.

Cherry trees do require light pruning in late spring to ensure even spacing of branches; get advice from your garden center or agricultural extension office, and never prune in winter—it can introduce disease.

⇾ PLUM ⇽

Prunus domestica
ROSACEAE (ROSE FAMILY)

When Americans think about a plum, we think about variations on the Japanese plum, *Prunus salicina*. These large, sweet, red or golden-fleshed plums were the invention of Luther Burbank, the most famous plant breeder of the twentieth century. From his farm in Santa Rosa, California, Burbank bred an astonishing eight hundred new varieties of plants, including the Shasta daisy, the Russet Burbank potato, and the Santa Rosa plum. In fact, almost all the plums grown in the United States today are Burbank's creations, hybrids of young trees he imported from Japan in 1887.

As wonderful as these plums are, we don't eat many of them. The average American eats just less than a pound of plums per year, and we use far fewer of them in booze. That is a tragedy that a few intrepid distillers are working to rectify.

European plums, *P. domestica,* have a long history in alcoholic beverages. There are over 950 varieties and many subspecies, all of which are subject to regular renaming and reclassification. The plums of most interest to the average drinker include four members of *P. domestica:* the bluish purple, oval-shaped damson (from *Damascus,* pointing to its ancient origins in Syria), the small golden mirabelle, the round bullace, which comes in a range of colors, and the pale lime-colored greengage. (The first three are usually assigned to the subspecies *insista,* while the gages are placed in a separate subspecies, *italica,* but even this is up for debate.) There are so many varieties of damsons, mirabelles, bullaces, and gages that even fruit growers can't keep them all straight; ask a farmer what variety of damson grows in the orchard and you might get nothing but a shrug in response.

But all of these plums make delightful liqueurs, eaux-de-vie, and brandies. The new Averell Damson Gin Liqueur from the American Gin Company, made with damsons grown in Geneva, New York, is the latest in a long line of fine damson liqueurs. Recipes for damson wine or damson-infused brandy date to 1717, and by the late nineteenth century, damson gin was a common drink in the English countryside. While it is a sweet liqueur, it is not cloying: modern, well-made damson gins are simply bright, clean expressions of a wild, natural plum flavor. Damsons, greengages, and bullace plums all grow wild in English hedgerows; both homemade and commercially produced liqueurs are made with each of them.

There is a bit of a botanical mystery surrounding the greengage. Many nineteenth-century botanical journals claim it was named after a member of the Gage family, who brought the tree back to England from the Chartreuse monastery sometime between 1725 and 1820, depending on which account you read. This anecdote is enough to send any inventive bartender scurrying around to create a cocktail that combines plum eau-de-vie and Chartreuse liqueur—but unfortunately, it is impossible to prove. An 1820 history of English fruits claims that some member of the aristocratic Gage family picked the trees up at the monastery and shipped them back to Hengrave Hall in Suffolk. Apparently a label was lost in transit and the French plum Reine Claude was simply labeled "Green Gage" to reflect the fruit's color and the estate where it was grown. Other accounts claim that a similar event took place within another branch of the Gage family on their estate called Firle.

All we know for certain is that "gage" plums were established in England before 1726, when they first appeared in horticultural literature, which means that if the Gage-Chartreuse label mixup happened, it surely took place well before 1725 or there wouldn't have been time for the trees to be planted, bear fruit, and attract the notice of horticulturists. An early mention of green plums, in a 1693 catalog of English plants, implies that an even earlier generation of Gages would have had to be involved. Arthur Simmonds, deputy secretary to the Royal Horticultural Society in the early 1900s, made a heroic effort to clear up the confusion, concluding only that the various Gage candidates put forth in the botanical literature were either not alive, old men, or young children when this mysterious trip to Chartreuse and subsequent label mixup occurred. Any connection between the Gage family and the green plums is, at this point, only speculative.

In France, the deep golden mirabelle plums are a specialty of the Lorraine region. In nearby Alsace the local plum is the quetsche, a violet-skinned fruit with yellowish green flesh. Each are made into jams, tarts, candies, liqueurs and remarkable eaux-de-vie. Eastern European countries are known for slivovitz, a kosher blue plum brandy that is often distilled with whole fruit and their pits, giving it a slight marzipan flavor, and sometimes aged in oak to add notes of vanilla and spice. While cheap imitation slivovitz, made from crude sugar-based liquor and prune juice, has a deservedly bad reputation, a well-made plum brandy or eau-de-vie is an extraordinary experience.

Other *Prunus* species are also put to use in liqueurs: for instance, Japanese plum wine, *umeshu,* is usually made with *P. mume,* a Chinese species more closely related to apricots. The ume fruit are soaked in a mixture of sugar and *shochu* (a spirit made of rice, buckwheat, or sweet potatoes bottled at 25 percent ABV) for up to a year before drinking. While commercial *umeshu* is available—sometimes with ume floating in the bottle—it is also something people make at home when the fruit is ripe.

QUANDONG

Santalum acuminatum

SANTALACEAE (SANDALWOOD FAMILY)

This Australian native is a hemiparasite, which means that it gets some, but not all, of its nutrition by robbing other plants. It thrives in poor soil, where its roots reach out to any nearby trees or shrubs, piercing their root systems and taking water, nitrogen, and other nutrients from them. It does produce its own sugars, but that's not enough to let it stand on its own. Quandong cannot grow unless other plants are nearby, making it difficult to cultivate.

The small red fruit is a uniquely Australian treat. Imagine a tart version of a peach, apricot, or guava. It's an aboriginal delicacy that has been made into jams, syrups, and pie fillings. The nuts were used as a traditional medicine; because they are encased in a hard shell, they would pass unmolested through the digestive tracts of emus and could be gathered from emu droppings.

But there's no need to go digging through emu droppings to enjoy a quandong cocktail. The fruits are being used by inventive Australian distillers eager to celebrate indigenous plants. Tamborine Mountain Distillery makes a quandong and gentian bitter liqueur; these and other products are helping to put quandong on fine cocktail menus around Australia.

ROWAN BERRY

Sorbus aucuparia
ROSACEAE (ROSE FAMILY)

Also called the European mountain ash, this flowering tree is not related to ash trees at all but is instead a relative of roses and blackberries. It thrives in hedgerows and wild areas throughout England and much of Europe, where the small, orange-red berries are prized for their high vitamin C content. They are used in homemade country wines, and to flavor traditional ales and liqueurs. An Austrian eau-de-vie called Vogelbeer, distilled from rowan berries, is an excellent example of an entire class of rowan berry spirits called *Vogelbeerschnaps.* Alsatian distillers, not to be outdone by the Austrians, make a lovely version of their own called *eau-de-vie de sorbier.*

SLOE BERRY

Prunus spinosa
ROSACEAE (ROSE FAMILY)

It took a renewed interest in wild, local, seasonal fruit to bring the sloe back from obscurity. Sloe gin, called snag gin in the nineteenth century, is nothing more than gin infused with sugar, perhaps some spices, and the small, astringent fruit of the thorny blackthorn shrub. It is a sweet red liqueur, much like damson gin, that people once made at home from fruit gathered

GROW *your* **OWN**

SLOES

Sloes are common hedgerow plants in England; in North America, they are difficult (but not impossible) to find at specialty fruit tree nurseries. These tough, hardy shrubs can form an impenetrable thicket given the opportunity. Expect them to reach fifteen feet and spread to at least five feet, but they can be pruned and kept small.

Plant sloes in full sun or light shade in moist, well-drained soil, preferably out of high-traffic areas as the thorns can be a nuisance. The shrubs are deciduous, meaning that they drop their leaves in winter, and will bloom in early spring and produce fruit in fall. Sloes are hardy to about −20 degrees Fahrenheit.

Leaving fruit on the branch until the first frost makes them a little sweeter, but the tart flavor is what makes them so good in sloe gin.

SHADE/SUN

REGULAR WATER

HARDY TO -20F/-29C

in the countryside. Syrupy, artificially flavored versions gave it a bad reputation in the twentieth century, but fresh ingredients and authentic recipes have returned. The makers of Plymouth Gin have come to the rescue, distributing their sloe gin internationally, and craft distillers are undoubtedly experimenting with sloes at this very moment.

The blackthorn is a close relative to the plum and cherry, but unlike those lovely trees, the blackthorn is not usually cultivated in orchards or gardens. It takes the form of a massive, fifteen foot-tall shrub covered in thorns and stiff branches. While it makes an excellent thicket or hedgerow, its messy habit and small, sour fruit make it the sort of plant that is best left in the countryside. It grows throughout England and most of Europe but is only cultivated in North America by the most dedicated growers of obscure fruit.

Its starry white flowers are among the first to appear in the spring, followed in fall by blackish purple fruits that can be harvested through the first frost. They are not sweet enough to be eaten on

SLOE GIN FIZZ

2 ounces sloe gin
½ ounce lemon juice (the juice of roughly half a lemon)
1 teaspoon simple syrup or sugar
1 fresh egg white
Club soda

Pour all the ingredients except club soda into a cocktail shaker without ice. Shake vigorously for at least 15 seconds. (This "dry shake" helps the egg white get frothy in the shaker; you can skip it if you'd rather not include egg whites.) Then add ice and shake for at least 10 to 15 seconds more. Pour into a highball glass filled with ice and top with club soda. Some people replace half the sloe gin with dry gin to make it less sweet, but try it this way first—you'll be surprised by how refreshingly tart it is.

their own, so sloes are made into jams and pies—but their highest and best use is sloe gin. The fruit is picked, washed, scored with a knife to break the skin, and soaked in gin or neutral grain spirits, along with sugar, for up to a year. The liqueur can be sipped neat— it's a nice winter pick-me-up—or mixed into a classic cocktail like the sloe gin fizz.

In the Basque region of Spain and southwestern France, a liqueur called *pacharán* or *patxaran* is made by macerating sloes in anisette, or a neutral spirit mixed with aniseed, and perhaps a few other spices such as vanilla and coffee beans. While it is commercially produced—Zoco is one such brand—families often make it themselves, and homemade versions are still served in small restaurants. Similar drinks include Germany's *Schlehenfeuer* and Italy's *bargnolino* or *prugnolino,* which combines sloes with a high-proof spirit, sugar, and either red or white wine. An *eau-de-vie de prunelle sauvage* is made in France's Alsace region.

Before sloe gin was adulterated with artificial flavors, it was itself an adulterant: added to bad wine, it passed for port in cheap wine shops. In their 1895 book *The New Forest: Its Traditions, Inhabitants and Customs,* authors Rose Champion De Crespigny and Horace Hutchinson noted that "when port wine went out of fashion we were told that it was made of log-wood and old boots. Since it has returned to fashion the demand for sloes has increased proportionately, affording strong grounds for inference that other things besides the log-wood and the boots are of its composition."

CITRUS

CITRUS: THE BARTENDER'S ORANGERIE

Citrus spp.

RUTACEAE (RUE FAMILY)

Imagine how difficult a bartender's job would be if every recipe involving citrus was taken away. Mojitos? Fresh lime is a requirement. Margaritas? They call for lime and triple sec, an orange liqueur. Martinis? Gin is flavored with citrus peels. Citrus adds a certain brightness, a certain sparkle, to most drinks. It boosts the top notes, those ephemeral floral and herbal flavors that might otherwise be lost in a complex distillation process. And remarkably, some of the most sour, inedible citrus make some of the best liqueurs.

Today's citrus varieties are the result of centuries of experimentation and hybridizing, making their exact lineage difficult to trace. All citrus trees we know today, including lemons and limes, probably originated from three unlikely candidates: the pomelo, a large, thick-skinned fruit like a grapefruit; the citron, with its formidable peel and distasteful fruit; and the sweet, thin-skinned mandarin. Some botanists believe that there were a couple other ancestors to modern citrus, which are now extinct.

Early records of citrus come from China, where four-thousand-year-old writings described people carrying bundles of small oranges and pomelos. Two thousand years later, the citron was moving across Europe. As impossible as it may be to imagine the Mediterranean and north Africa without its citrus trees, Arab traders brought the sour orange, the lime, and the pomelo to the region only eight hundred to a thousand years ago. The sweet orange came only four hundred years ago, when Portuguese traders carried it back from China. By this time, citrus was moving all over the world—sometimes with strange and unintended consequences.

On Columbus's second voyage to the Americas, in 1493, he brought sweet oranges with him and made some attempts to establish them in the Caribbean. Just a few decades later, the first orange trees showed up in Florida. But something surprising happened when explorers, familiar only with growing conditions in their Mediterranean climate back home, planted citrus in the hot, tropical Caribbean region.

First, many of the trees refused to produce orange fruit. In the hottest weather, citrus can remain stubbornly green. It turns out that the most vivid color develops only when the night air takes on a slight chill, as it does in California or, for that matter, in Spain or Italy. Cooler temperatures break down the chlorophyll in the rind, letting the orange pigments show through. In hot climates, the fruit might taste sweet, but the rind remains tinged in green and yellow.

The other surprise? Some trees became mutant freaks after being planted on tropical islands, producing bitter, pithy fruit with thick rinds that seemed to have no culinary value. But colonists, desperate to make use of the crops they'd worked so hard to plant, discovered that drowning them in hard liquor improved them considerably.

⊰ BITTER ORANGE ⊱

Citrus aurantium

This sour orange, also called the Seville bitter orange, came to Spain by way of the Moors in the eighth century. It was probably never eaten raw, but its peel quickly found its way into liqueurs, perfume, and marmalade. As dreadful as the juice might be on its own, it is the essential ingredient in *mojo,* a marinade combining the bitter orange juice with herbs and garlic.

Bitter oranges also flavor triple sec. Although many orange liqueurs go by the name triple sec, a French distiller, Combier, lays claim to the original recipe. They offer up a royal legend to explain their elixir's origin: the company's story involves a chemist, François Raspail, who was imprisoned after he ran unsuccessfully against Napoleon III and later led a revolt against him. Raspail was also a noted botanist—he was one of the first to use microscopes to identify plant cells—and he'd apparently created a medicinal potion from aromatic plants. In prison, the story goes, he met confectioner Jean-Baptiste Combier, who was also serving time for denouncing Napoleon III's authoritarian rule. Combier had already developed an orange liqueur recipe with his wife. The men agreed that when they were released, they'd go into business together, combine their recipes, and release the result as Royal Combier.

Imprisoned chemists aside, what modern drinkers need to know is that triple sec as it is made by Combier is a sugar beet spirit combined with bitter orange peel. Even this high-quality version is not complex enough to be drinkable on its own; every good triple sec tastes more or less like orange candy. It is nonetheless worth seeking out a quality orange liqueur for margaritas, sidecars, and other recipes that call for it.

We have Spanish explorers to thank for bringing their bitter Seville oranges to Curaçao, an island of the Lesser Antilles off the coast of Venezuela. The variety that grew from those early discarded seeds came to be called Laraha (*Citrus aurantium* var. *curassaviensis*). They tasted terrible, but desperate sailors ate them anyway as a treatment for scurvy after a long journey across the ocean. In fact, the island's name may come from the Portuguese word for "cured."

And, of course, Laraha was made into liqueur. Originally the peels were dried in the sun and soaked in spirits along with other spices. Today, according to the makers of the real curaçao liqueur, there is still an original plantation of forty-five Laraha trees on the island. Twice a year the trees are harvested, producing only nine hundred oranges. After drying the peels in the sun for five days, they are suspended inside the still in jute bags to extract the citrus flavors. Then other flavors are added—the exact recipe is a secret, but nutmeg, clove, coriander, and cinnamon are likely suspects—and it is bottled with or without food coloring. Curaçao is known for its vivid, Caribbean blue color, but this is simply an artificial color and true curaçao can be purchased without it.

The extract of bitter orange can also be found in Grand Marnier, a Cognac-based liqueur. The peels are left to dry in the sun and then soaked in a high-proof neutral alcohol to extract the flavor. That essence is then combined with Cognac and a few other secret ingredients, then aged in oak. Grand Marnier works as a mixer for any cocktail that calls for a citrus liqueur, giving it a rich elegance that other orange liqueurs lack.

AND NOW FOR A BOTANICAL QUIBBLE

The distillers of Grand Marnier claim to flavor it with the rinds of a fruit called *Citrus bigaradia*, but don't try to find it at a nursery: the name dates back to 1819 but is no longer in use by botanists. At best it refers to a particular variety of the *C. aurantium* species, *Citrus × aurantium* var. *bigaradia*.

RED LION HYBRID

This variation on the classic Red Lion is designed to showcase the flavor of Grand Marnier, like the original did, but also to feature fresh, seasonal orange juice. It's fantastic in the winter when tangerines are at their peak.

1 ounce Plymouth gin or vodka
1 ounce Grand Marnier
¾ ounce freshly squeezed orange or tangerine juice
Juice of one freshly squeezed lemon wedge
Dash of grenadine
Orange peel

Shake all the ingredients except the orange peel over ice and serve in a cocktail glass. Garnish with the orange peel.

WHAT'S BEEN SPRAYED ON THAT ORANGE ZEST?

IN FLORIDA AND TEXAS, AND ON WARM CARIBBEAN ISLANDS, citrus groves don't experience the cool nights needed to help the fruit turn from green to orange. This forces growers to find other ways to use greenish fruit that is perfectly ripe but unattractive. This problem explains, in part, why Florida is so well known for its juice industry while California, which has cooler nights, sells more fresh citrus. Some growers correct the green color by exposing the fruit to ethylene, a naturally occurring gas that speeds up ripening and breaks down chlorophyll.

Farmers in the United States are also permitted to spray the fruit with a synthetic dye called Citrus Red No. 2. The dye is banned in California but may be used by Texas and Florida growers. It is only permitted for fruit that is going to be peeled and eaten or juiced, not for fruit whose rinds will be "processed" into food or drink. Because fruit sold at the grocery store is assumed to be for eating or juicing, it may be sprayed with the dye—and not always labeled as such.

Citrus can also be sprayed with wax; if wax is used on organic citrus, it cannot be synthetic or petroleum-based. If you'd prefer to avoid using synthetic dyes or waxes in cocktails, *limoncello*, or other infusions, choose organic citrus.

ESSENTIAL OILS

An essential oil is a volatile oil extracted from a plant through distillation, pressing (expression), or solvents. In the case of citrus, the most common oils are:

NEROLI OIL	Extracted from the blossoms of bitter orange, usually through water distillation
PETITGRAIN OIL	A distillation of the leaves and twigs of a citrus tree
SWEET ORANGE OIL	Extracted from the rind of the orange, often through cold pressing

⊹ CALAMONDIN ⊹

Citrofortunella microcarpa (syn. *Citrus microcarpa*)

A likely cross between a mandarin and a kumquat, the calamondin retains the best qualities of both trees: small fruit with thin skin and a tart but not bitter juice. It is one of the most cold-tolerant of all citrus trees, surviving even when the temperature dips below freezing, and it is so content in a pot indoors that it has become popular as a houseplant. It is widely grown in the Philippines, where it is also called *calamansi*.

The juice is just tart enough to substitute for a lime in cocktails. The peels can be soaked in vodka and sugar to be made into a liqueur. In the Philippines, the juice is treated as a mixer with vodka and club soda.

⊰ CHINOTTO ⊱

Citrus aurantium var. myrtifolia

With diminutive fruit no larger than a golf ball and tiny, diamond-shaped leaves, the chinotto (pronounced key-NO-toe) is the kind of tree that any citrus collector would want in an orangerie. Although the fruit is often described as bitter and sour, it's actually less tart than a lime or lemon and perfectly fine to eat. The trees flourish in the Mediterranean, where the fruits ripen in January.

The chinotto's distinctive flavor is widely reported to be a key ingredient in Campari, which is best enjoyed in a Negroni or splashed into club soda. A nonalcoholic soda called Chinotto can also be found throughout Italy and in Italian markets everywhere. Resist the temptation to combine the two: mixing Campari and Chinotto would almost certainly be too much of a good thing.

NEGRONI

1 ounce gin
1 ounce sweet vermouth
1 ounce Campari
Orange peel

Shake all of the ingredients except the orange peel over ice and serve in a cocktail glass. Garnish with the orange peel.

⊰ CITRON ⊱

Citrus medica

One of the earliest species of citrus and the parent to many others, the citron is known for its monstrously thick peel and sour, nearly inedible fruit. In about 30 BC, Virgil wrote that citron "has a persistently wretched taste, but is an excellent remedy against poisons." The peel was added to wine as a medicinal remedy; it induced vomiting, which might not recommend it as a cocktail ingredient.

Citron is the dinosaur of the citrus world. It is downright reptilian in appearance, with thick, wrinkled skin and bizarre deformities. The Buddha's hand citron (*Citrus medica* var. *sarcodactylis*) is shaped like a many-fingered hand, making it almost all peel and no flesh. Like other citron, it can be brined and candied to make into a kind of crystallized peel called succade. But because the Buddha's hand has such a large surface area of flavorful peel, it can also be infused whole in vodka.

Recipes for "citron water" from Barbados, where the tree was abundant, date to before 1750 and might have flavored vermouth in those days. The fruit was also chopped or zested, soaked in a variety of spirits, and mixed with sugar to make a cordial not unlike *limoncello*.

CITRUS PEEL: THE RIGHT TOOL FOR THE JOB

The best tool for peeling citrus is a handheld zester, which looks like a fat, stubby fork. The tines on the end are used to extract zest, but below the tines is a hole with a sharp edge that makes long, thin, perfect peels.

GROW *your* OWN

CITRUS

Anyone who lives in a mild-winter climate and doesn't have a citrus tree in the backyard is squandering an opportunity. There is nothing better than grabbing a fresh lemon or lime for a cocktail, and even neglected trees producing nearly inedible fruit still offer excellent peel for garnishes.

If possible, visit a fruit tree nursery that specializes in citrus to choose a tree with fruit you like that grows well in your area. At a general garden center, ask around and find an employee with expertise in citrus who can advise you about potential pests or disease in your area and tell you whether young trees will need protection against frost.

Calamondin, Improved Meyer lemon, and most lime trees do well in containers and can survive indoors if they have bright light (not just a sunny window, but a well-lit conservatory, greenhouse, or supplemental grow lights) and if you can keep their living quarters humid in winter, when furnaces make the air too dry for their taste. Potted citrus should be kept on the dry side in winter, as cold, wet roots may rot.

Use specialized citrus fertilizer monthly during the growing season, but withhold it in the winter, when it can burn roots already stressed by cold temperatures. Nearly all citrus are self-fruitful, meaning that they don't need another tree nearby for pollination.

FULL SUN

REGULAR WATER

HARDY TO 30F/-1C

⊰ GRAPEFRUIT ⊱

Citrus × paradisi

A cross between the sweet orange and pomelo, grapefruit is most likely a mutant, or an accidental hybrid, that sprang up around 1790 on Barbados. The compelling mixture of tangy citrus and bitterness make grapefruit an astonishingly good mixer—it works well in variations of the Negroni, and it blends beautifully with either rum or tequila.

Grapefruit liqueurs are harder to come by. Giffard Pamplemousse is one example; it is made from a maceration of pink grapefruit. An Argentinian distillery called Tapaus makes Licor de Pomelo, *pomelo* being the Spanish word for "grapefruit." Both could be sipped on their own or used experimentally in any cocktail that called for citrus liqueur.

CIAO BELLA (A NEGRONI VARIATION)

1 ounce **gin**
1 ounce **sweet vermouth**
1 ounce **Campari**
1 ounce **grapefruit juice**
Grapefruit zest

Shake all of the ingredients except the grapefruit zest over ice and serve in a cocktail glass. Garnish with a wide slice of grapefruit zest.

ICHANG PAPEDA (C. ICHANGENSIS): THE WORLD'S HARDIEST EVERGREEN CITRUS, SURVIVING TEMPERATURES DOWN TO 0 DEGREES FAHRENHEIT IN THE HIMALAYAN FOOTHILLS. THE FRUIT OFTEN CONTAINS NO JUICE AT ALL, JUST SEEDS AND PITH, MAKING IT FRAGRANT BUT NEARLY INEDIBLE.

FRANK N. MEYER, PLANT HUNTER

JAPANESE IMMIGRANTS BEGAN IMPORTING SWEET LEMON-MANDARIN CROSSES TO THE UNITED STATES IN THE 1880S, but the Meyer lemon bears the name of the man responsible for formally introducing it to America. Frank N. Meyer was born in Amsterdam in 1875 and arrived in New York City in 1901. He undertook four expeditions to Russia, China, and Europe on behalf of the USDA, gathering seeds and plants that might be of use to American farmers. In all, he introduced twenty-five hundred new plants, including the Chinese persimmon, the gingko tree, and an astonishing array of grains, fruits, and vegetables. He also endured unimaginable hardship, including injuries, illnesses, robberies, and the loss of countless plant specimens owing to shipping problems or delays clearing customs.

He found what is now the Meyer lemon in Peking in 1908 and managed to get it back to the States. Over the next few decades, farmers realized that clones of this tree were symptomless carriers of a disease called tristeza; as a result, many of the original Meyer lemons had to be destroyed. A virus-free selection was discovered by Four Winds Growers, a California nursery, in the 1950s. Today the Improved Meyer lemon is once again widely grown.

Mr. Meyer's plant explorations came to a tragic end in 1918, when at the age of forty-three he died while traveling down the Yangtze River to Shanghai. His body was recovered from the river a week later, though the exact cause of death remained a mystery.

⊰ LEMON ⊱

Citrus limon

A lemon is most likely a cross between a lime, a citron, and a pomelo. The Italian Sorrento lemon, Femminello Ovale, definitely exhibits citron characteristics, with its thick skin and sour flavor.

To get the flavor just right, Sorrento trees are shaded by straw mats called *pagliarelle* or, more recently, plastic shade cloths. This protects the trees from cold weather and helps slow down the ripening process so that harvest season happens in summer. Because the trees produce fruit year-round, each crop has its own name: *limoni* comes first, in the winter, followed by *bianchetti*, then *verdelli* during the summer months and *primofiori* in the fall.

The Eureka lemon, more properly called Garey's Eureka, is descended from the Sicilian lemon and is a more acidic, thick-skinned variety. The most popular lemon for home gardeners, cooks, and bartenders is the sweet and juicy Meyer lemon, which is actually a cross between a lemon and a mandarin. The rind is lower in essential oils, so for mixing drinks, the zest is less desirable than the juice itself.

THE FRANK MEYER EXPEDITION

This combination of straight spirits, sugar, and Meyer lemon showcases the fruit perfectly. The Champagne float gives it a nice effervescence. Mix up a batch for friends and drink a toast to Mr. Meyer and his daring adventures.

1½ ounces **vodka**
¾ ounce **simple syrup**
¾ ounce **Meyer lemon juice**
Dry sparkling **wine (Spanish *cava* works well) or sparkling water**
Lemon **peel**

Shake the vodka, simple syrup, and lemon juice over ice and strain into a cocktail glass. Float sparkling wine on top and garnish with lemon peel. For a less intoxicating variation, strain into a tumbler over ice and top with sparkling water instead of sparkling wine.

⊰ LIME ⊱

Bearss lime, Tahiti lime, or Persian lime, *Citrus latifolia*
Key lime, Mexican lime, or West Indian lime, *Citrus aurantifolia*
Kaffir lime, *Citrus hystrix*

Limes originated in India or Southeast Asia and came to Europe by the fifteenth century. They are actually yellowish green when ripe; they have to be picked before ripening to retain the green color that buyers expect in limes. With half the sugar content of lemons, and slightly more acid, they play a distinct role in cocktails. Chemical analysis of limes shows that they are higher in linalool and α-terpineol, two rich, floral flavors, and that the peel contains oils that add a warm, spicy note.

The more acidic key lime is the bartender's best friend, adding just the right tropical touch to margaritas and mojitos. It also grows particularly well in a container, staying small and producing fruit nearly year-round. The milder Bearss lime, considered the "true lime," produces larger fruit and tolerates cooler climates. The kaffir is grown primarily for its leaves, which flavor Thai food and are used in infused vodkas. Its rind is grated into curries, but the fruit itself is nearly inedible.

A number of lime liqueurs are on the market, the most useful being Velvet Falernum, made with lime, sugar, and spices. (There are also nonalcoholic lime, spice, and sugar mixers sold as Falernum that accomplish the same thing in a drink.) Mai tais, zombies, and other tropical cocktails depend upon falernum. A French liqueur called Monin Original Lime, which was introduced in 1912, has only recently been back on the market and is hard to find in the United States, but it is worth seeking out for citrus-based drinks. St. George Spirits makes a kaffir-infused Hangar One vodka; it is the perfect base to Thai-inspired cocktails.

FLAVEDO OR EXOCARP The zest, which contains oil glands, fatty acids, flavor, enzymes, pigments, and a bitter aromatic compound called limonene.

ALBEDO OR MESOCARP The pith, a spongy white layer that is usually not eaten, although it contains healthy phytochemicals. The term *pith* also applies to the thin stringy membrane that clings to the edible segments.

ENDOCARP The inner layer that directly surrounds the seeds. In the case of citrus, this is the part that is eaten. (In other fruit, such as peaches, the mesocarp is eaten and the endocarp is merely a thick, fibrous membrane clinging to the pit.)

❄ MANDARIN ❄

Tangerine, clementine, or common mandarin, *Citrus reticulata*

Chinese mandarin, *C. nobilis*

Satsuma mandarin, *C. unshiu* (syn. *C. reticulata*)

The much-hybridized mandarin, a sweet fall or winter-fruiting orange with a loose skin that practically falls off the fruit, flavors a Cognac-based liqueur called Mandarine Napoleon that, according to its makers, has its origins in Napoleon's court. Chemist Antoine François, comte de Fourcroy, apparently invented the recipe for Napoleon, who liked his brandy steeped in orange peels. And in fact, mandarins grew in Corsica, an island off the coast of northern Italy, which was the French emperor's birthplace. Mandarin blossoms, along with a little peel, also flavor a delightful mandarin-infused Hangar One vodka from St. George Spirits.

❄ POMELO ❄

Citrus maxima (syn. *C. grandis*)

Also called shaddock, the pomelo is an ancestor to modern grape-fruits and bitter oranges. The fruit is large and heavy, weighing up to four pounds. The rind has a thick, often green skin, especially in southeast Asia, where it is widely grown.

Charles Jacquin et Cie, makers of Chambord raspberry liqueur, once made a pomelo and honey brandy–based liqueur called Forbidden Fruit that was an essential ingredient in some classic cocktails, including the Tantalus, a mixture of equal parts lemon juice, Forbidden Fruit, and brandy. (Some bartenders attempt to re-create the liqueur by steeping pomelo or grapefruit peel, honey, spices, and vanilla in brandy, with varying degrees of success.) The words *pomelo* and *pummelo* are widely used to refer to either true pomelos or grapefruits, so liqueurs with *pomelo* in the name might be flavored with either fruit.

⊰ SWEET ORANGE ⊱

Citrus sinensis

The sweet orange, probably a cross between pomelo and mandarin, is one of the most widely grown fruit trees in the world, accounting for almost three-quarters of all citrus production. Valencia, Navel, and blood oranges are the best-known varieties. While they are popular for fresh fruit and juice, they are not the top choice for distillers making citrus-flavored liqueurs. Those liqueurs tend to be flavored with the more complex, bitter sour oranges. However, the peel is widely available through spice distributors, so it is often used to add a bright note to gins and herbal liqueurs.

One orange-flavored liqueur that employs sweet oranges is Orangerie, which its distiller describes as a blend of hand-zested Navalino oranges (botanists do not recognize a variety called Navalino, but perhaps they mean Navelina, a sweet navel orange from Spain first described in 1910), cinnamon, and cloves, infused in Scotch whisky. Another is Solerno Blood Orange Liqueur, a sweet liqueur made from Sanguinello blood oranges that combines separate distillations of the fruit, the peel, and lemon rind. It's an upscale substitute for triple sec and adds a lively, sweet note to gin drinks.

BLOOD ORANGE SIDECAR

This variation on the classic sidecar replaces lemon juice with blood orange juice. Feel free to adjust the proportions to taste. And if you're not a fan of brandy, replace it with bourbon. (And if you're not a fan of bourbon, go read another book. No, seriously, experiment with your spirit of choice. Vodka, gin, rum? Give it a try!)

1½ ounces **Cognac or brandy**
¾ ounce **blood orange juice**
½ ounce **Solerno Blood Orange Liqueur**
    (or another citrus liqueur like triple sec)
Dash of **Angostura bitters**

Shake all the ingredients except the bitters over ice and strain into a cocktail glass. Add a dash of bitters on top.

YUZU

Citrus × junos (syn. *C. ichangensis × C. reticulata* var. *austere*)

This thick-skinned, sour cross between a mandarin and the strange, bitter Ichang Papeda comes from China and made its way to Japan around 600 AD. While the fruit is not particularly tasty, the rind exudes a complex, fruity citrus fragrance beloved by Japanese cooks. *Yuzu* zest can be found in a soy sauce called *ponzu,* and it also flavors miso soup. People bathe with it as well; a traditional Japanese solstice bath features *yuzu* fruit floating in hot water.

Yuzu is an enchanting addition to flavored sake and *shochu*-based liqueur. A Korean *yuzu* syrup called *yucheong,* available at Asian grocery stores, is mixed with hot water to make tea but also happens to be a fantastic cocktail ingredient.

Because *yuzu* trees are hardy to 10 degrees Fahrenheit, they survive in mountainous regions where no other citrus can be found. Gardeners in England and colder regions of the United States who are determined to grow citrus outdoors might have luck with *yuzu* if other citrus have failed to thrive.

RAMOS GIN FIZZ

New Orleans bartender Henry Ramos is credited with inventing this drink around 1888. During the 1915 Mardi Gras Carnival season, he created quite a spectacle by having thirty-five muscular bartenders line up and shake the drinks. Many bars won't make it, fearing the liability of serving raw eggs, or dreading the effort involved. At Graphic, London's excellent gin bar, it is not unusual for a Ramos Gin Fizz to be passed around the room and shaken by bartenders, waitresses, and customers until the froth is perfect.

1½ ounces gin (the original recipe called for Old Tom gin)
½ ounce lemon juice
½ ounce lime juice
½ ounce simple syrup
1 ounce cream
1 egg white
2 to 3 drops orange flower water
1 to 2 ounces soda water

Combine all the ingredients except the soda water in a cocktail shaker and shake without ice for at least 30 seconds. Then add ice and continue to shake for at least 2 minutes, passing the shaker around the room as needed to keep it going without incurring frostbite. Pour the soda water into a highball glass and strain the fizz into the glass.

ORGEAT (PRONOUNCED OR-ZHA, ALTHOUGH MANY AMERICANS PRONOUNCE IT OR-ZHAT): A SWEET, OFTEN NONALCOHOLIC SYRUP MADE WITH ALMONDS, SUGAR, AND ORANGE FLOWER WATER, SOMETIMES IN A BASE OF BARLEY WATER. ORGEAT IS AN ESSENTIAL INGREDIENT IN A MAI TAI, ALTHOUGH IT IS LEFT OUT ALL TOO OFTEN.

MAI TAI

1½ ounces dark rum (some recipes mix dark and light rum)
½ ounce lime juice
½ ounce curaçao or another orange liqueur
Dash of simple syrup
Dash of orgeat syrup
Maraschino cherry
Wedge of pineapple

Shake all the liquid ingredients and strain. Serve over crushed ice in a goblet or highball glass. Garnish with the cherry and a pineapple wedge. If you have ever been tempted to put a paper umbrella in a glass, this would be the time.

ORANGE LIQUEURS: A PRIMER

LIQUEUR	BASE SPIRIT	INGREDIENTS	AGED IN OAK?
Cointreau	Sugar beet	Sweet and bitter orange peel	No
Combier	Sugar beet (Royal Combier also includes Cognac)	Bitter Haitian and sweet Valencia oranges	No
Senior Curaçao of Curaçao	Sugar cane	Laraha oranges	No
Grand Marnier	Cognac	Bitter orange, vanilla, spices	Yes
Mandarine Napoleon	Cognac	Dried mandarin peel, herbs, spices	Yes
Orangerie	Scotch whisky	Orange peel, cinnamon, cloves	Yes
Solerno blood orange liqueur	Neutral spirits	Blood orange fruit, zest, and Sicilian lemons	No
Generic triple sec or curaçao	Varies by distiller, usually neutral grain, sugar beet, cane sugar, or grape spirits	Sweet and bitter orange peel	No

-- and wrapping things up with --
NUTS & SEEDS

nut:

A DRY FRUIT THAT DOES NOT
OPEN AT MATURITY TO RELEASE ITS SEED;
GENERALLY SURROUNDED BY
A HARD WOODY OUTER COVERING
AND CONTAINING ONLY ONE SEED.

seed:

A STRUCTURE CONTAINING
AN EMBRYO WHICH FORMS IN A PLANT'S
OVARY FOLLOWING FERTILIZATION.

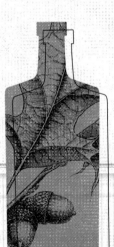

ALMOND

Prunus dulcis

ROSACEAE (ROSE FAMILY)

T here is drawne out of sweet Almonds, with liquor added, a white juice like milke." So said John Gerard, English barber-surgeon and herbalist who in 1597 published *The Herball, or Generall Historie of Plantes,* a vivid but fanciful compendium of botanical knowledge and half-truths. He claimed that chestnuts could keep horses from coughing and that the juice of basil leaves could treat snakebite—but he did get a few things right. Sweet almonds? Liquor? Gerard was on to something.

Almonds are quite closely related to apricots and peaches and probably share an Asian heritage. The trees were cultivated in China twelve thousand years ago and made their way to Greece by the fifth century BC. They prefer a Mediterranean climate, with mild winters and long, dry summers, which helped them to spread successfully across Asia and into southern Europe, northern Africa, and the west coast of the United States. They are so abundant in California that hives of European honeybees are carried from orchard to orchard to pollinate the crop.

The nuts weren't always so delightful to eat. Bitter almonds, *Prunus dulcis* var. *amara,* contain enough cyanide to be deadly at a dose of fifty to seventy nuts. Fortunately, people are unlikely to eat a bitter almond by mistake; they are not sold in stores and are grown primarily for pressing into almond oil, using a process that removes the poisons.

It is the sweet almond, *Prunus dulcis* var. *dulcis,* that lends its unmistakable honeyed nuttiness to liqueurs. The poisons have been bred out of this variety through centuries of selection, with orchardists choosing trees that happened to produce sweeter, less toxic almonds.

Almond liqueurs have been popular since the Renaissance, an era of great discovery that included the realization that any number of wonderful things happen when fruit, spices, and nuts are soaked in brandy. The goal could have been to create a medicine, or simply to soften the edges of a crudely distilled spirit. The Italian amaretto is the best-known example, although the brand most widely sold around the world, Amaretto di Saronno, contains no almonds at all but instead gets its nuttiness from the kernels of a close botanical relative, the apricot. Still, it is fairly easy to find an amaretto made with actual almonds: try Luxardo Amaretto di Saschira Liqueur.

Although the liqueur is perfect on its own, it is also used to flavor biscotti. There are few better ways to end a meal than with amaretto-laced coffee and biscotti.

AN ALMOND IS NOT TECHNICALLY A NUT. FROM A BOTANICAL PERSPECTIVE, A NUT IS A FRUIT WITH A DRY, HARD SHELL. AN ALMOND IS A DRUPE, OR A STONE FRUIT WHOSE PIT SURROUNDS A FLESHY SEED. UNLIKE PEACHES, APRICOTS, AND OTHER DRUPES, HOWEVER, THE ALMOND'S "FRUIT" IS NOTHING MORE THAN AN UNAPPETIZING LEATHERY OUTER MEMBRANE.

COFFEE

Coffea arabica
RUBIACEAE (MADDER FAMILY)

What we refer to as a coffee bean is actually a pair of seeds found inside a small, red fruit: the coffee "cherry." The fruit grows on an Ethiopian shrub that claims both quinine and gentian as relatives. (All are in the taxonomic order Gentianales.) It produces a remarkable poison that will paralyze or kill an insect attempting to feed on it. That poison, caffeine, is exactly what drew us to the plant seven hundred years ago. Humans are not immune to the poison, but it would take over fifty cups of coffee, downed in rapid succession, to deliver a fatal dose.

Arab traders first brought coffee from its native Africa to Europe sometime before 1500. It took over a century for it to catch on, but by the mid-1600s, coffee houses were well established in England and throughout Europe. A charming story circulated about an Ethiopian goatherd whose goats ate the fruit of a coffee shrub and were so energized that they jumped and frolicked for the rest of the day and didn't sleep that night. Although this was probably nothing but a tall tale recited by merchants, it persisted well into the nineteenth century. The fact that a plant could allow people to go without sleep was considered a major scientific breakthrough.

In the early 1700s, Dutch and French traders took just a few varieties of coffee to plantations in the Americas, inadvertently creating a sort of genetic bottleneck. A surprising lack of diversity among coffee plants continues today. Although there are over a hundred known species, almost all coffee grown around the world comes from clones of *Coffea arabica,* with *C. canephora* (sometimes called *C. robusta*) in second place. Insect and pest problems among this monocrop have sent botanists in search of other species, some of which are near extinction in their native habitat. Plant explorers from Kew Gardens have discovered thirty previously unknown species of coffee in the last decade, each with their own remarkable characteristics: some contain almost no caffeine, others produce seeds twice the size of anything seen before, and some, it is hoped, will be better able to resist pests and disease.

There is no easy way to harvest coffee. It has to be handpicked because the fruit does not all ripen at the same time. The green seeds must be separated from the fruit, which can be accomplished through a "wet" process, in which the seeds are picked out of the fruit and fermented in water to remove pulpy residue, or a "dry" process in which the fruit is dried so that it can be more easily separated from the seeds. (The wet process is believed to produce a better-tasting bean and commands a higher price.) Once the green seeds are clean, they're ready to be roasted.

Coffee is now grown in fifty countries and has surpassed tea as the global drink of choice: we produce three times more coffee than we do tea. But learning to grind it and boil it in water was only the first step. By the early 1800s, coffee was being made into liqueur as well. Most early recipes called for nothing more than roasted coffee beans, sugar, and some sort of spirit. Such a product was produced commercially by 1862, when it was shown at the International Exhibition in London. Early twentieth-century recipes added cinnamon, cloves, mace, and vanilla.

By the 1950s, the rum-based Mexican liqueur Kahlúa was gaining popularity. Unlike many liqueur companies, this one does not keep the recipe a secret: the sugarcane spirit is barrel-aged for seven years, then combined with coffee extract, vanilla, and caramel. Dozens of coffee liqueurs are now sold around the world, based on spirits ranging from rum to Cognac to tequila. Craft distillers are partnering with specialty roasters to create high-end coffee spirits. Firefly in Santa Cruz, California, is one such example. They blend wet-processed Costa Rican beans with a brandy made from Syrah and Zinfandel grapes. Bartenders are also doing their own coffee infusions behind the bar, muddling beans into cocktails, and using coffee bitters in spicy drinks.

But perhaps the best-known combination of coffee beans and alcohol is Irish coffee. As with most famous drinks, its history is hotly debated, but one version of the story credits a bartender in Ireland's Shannon Airport as the inventor. A travel writer returning from Ireland asked a bartender at the Buena Vista restaurant in San Francisco to re-create it, and after much experimenting, the perfect combination of coffee, whiskey, sugar, and cream came together in the glass.

BUENA VISTA'S IRISH COFFEE

Hot coffee
2 sugar cubes
1½ ounces Irish whiskey
2 to 3 ounces whipping cream, lightly whipped with a whisk

Fill a heat-resistant glass or mug with hot water to warm it. Empty it and pour in the coffee until it is two-thirds full. Add the sugar cubes and stir vigorously; then add the whiskey. Carefully top with whipped cream.

HAZELNUT

Corylus avellana

BETULACEAE (BIRCH FAMILY)

T he hazel tree traces its origins to Asia and parts of Europe, where it has been actively cultivated for over two thousand years. The French gave the nut the name filbert, presumably after seventh-century abbot St. Philibert, whose feast day is August 20, precisely when the nuts are ripe. But the English called it a hazelnut. Over time, botanists settled the disagreement by assigning the word *filbert* to one species, *Corylus maxima,* and *hazelnut* to another, *C. avellana.* In the United States, the two words are used interchangeably, much to everyone's confusion, even though most farmers grow *C. avellana.* There are native American species, but they're not as productive as the European trees.

Although hazel trees can reach fifty feet in height, they tend to be short and shrubby, and farmers encourage that behavior. They lend themselves to coppicing, a practice of cutting down the main trunk of a tree to encourage twiggy growth from the roots. This keeps them productive and makes the harvest easier to manage.

Roasted hazelnuts in particular have a sweet, caramelized flavor that comes from at least seventy-nine different flavor compounds. Raw nuts have fewer than half that number, so the roasting process is vital to bringing out their complex taste.

Hazelnut liqueurs like Frangelico and Fratello are sweet blends of hazelnut and other spices like vanilla and chocolate. The Frangelico distillery crushes its toasted hazelnuts and then extracts the flavors in a mixture of water and alcohol. Some of this infusion is distilled, so that the final version contains both the distillate and the infusion. Also added are vanilla, cocoa, and other extracts.

That is the Italian style; the French version would be something more like Edmond Briottet's Crème de Noisette, a pale amber liqueur with a bright, clear hazelnut flavor. Craft distilleries in the Pacific Northwest have also begun experimenting with hazelnut-infused vodka and hazelnut liqueurs. And behind the bar, hazelnuts are an ingredient in small-batch bitters, and pure hazelnut extract can be used as a cocktail ingredient or whipped into cream for nutty coffee drinks.

KOLA NUT

Cola acuminata
STERCULIACEAE (CACAO FAMILY)

This African tree, a relative of the South American chocolate-producing cacao, grows to over sixty feet tall in its natural state and unfurls sprays of exquisite pale yellow flowers streaked in purple. After it blooms, a cluster of leathery, wrinkled fruit emerges, each containing about a dozen seeds. Those seeds are the kola nut, a mildly caffeinated treat enjoyed by West Africans as a stimulant. Once Europeans discovered it, the nut followed a now-predictable journey from eighteenth century medicine to nineteenth century tonic to twentieth-century flavoring extract.

Kola elixirs were prescribed for seasickness and as an appetite stimulant, often in combination with gentian and quinine. Early recipes for kola bitters were straightforward combinations of kola nut, alcohol, sugar, and citrus. By the late 1800s, kola wine and kola bitters were available in markets in London, and French and Italian distillers were releasing aromatized wine and *amaro*s with kola as an ingredient. Toni-Kola, an aperitif wine, is one famous and now defunct brand.

Soda fountains stocked kola syrup to make cocktaillike mixtures of fizzy, nonalcoholic drinks in the early twentieth century; these elaborate drinks were seen as one way to encourage temperance. The Coca-Cola company fought endless trademark battles over its use of the word "cola" to describe its products, but courts remained firm that "cola" was a general term to describe any beverage made with an extract of the kola nut and could not be trademarked. It remains an approved food flavoring today, and many natural soda companies still use the nut to add caffeine and that sweet, round cola flavor.

South Africans can buy a sweet syrup called Rose's Kola Tonic, and British, Australian, and New Zealand drinkers can look for Claytons Kola Tonic, a mixer that is also marketed as something a nondrinker can order in a bar (much like any other cola). Master of Malt, a UK liquor retailer, sells kola bitters in a dark rum base, which they promise will deliver "depth, tang and astringency" to cocktails. And although Italian *amaros*, including Averna Amaro and Vecchio Amaro del Capo, are described as having "notes of cola," the manufacturers offer no clue as to whether the nut is in fact part of their secret formulas.

WALNUT

Juglans regia

JUGLANDACEAE (WALNUT FAMILY)

T here is nothing as astringent and unpleasant as a green, unripe walnut—until it has been soaked in alcohol and sugar, that is. *Nocino,* an Italian walnut liqueur, is surely one of the most ingenious uses of surplus produce ever invented.

Walnut trees are native to China and eastern Europe, and still grow wild in Kyrgyzstani forests. They were introduced to the West Coast by Franciscan monks around 1769, and are still seen growing on the grounds of California missions. The black walnut, *J. nigra,* is native to the eastern United States and is prized as much for its durable, dark wood as its fruit. Because it tolerates cold temperatures so well, European explorers brought the black walnut to Europe in the seventeenth century.

The magnificent trees reach over one hundred feet and cast a wide shadow. Long, ropy clusters of male flowers, called catkins, emerge in spring and release pollen, which is captured by the decidedly unglamorous green female blossoms. A soft green fruit emerges after pollination, and by early summer, the tree is laden with more walnuts than it can possibly support. Many of them drop to the ground before autumn.

This must have frustrated early orchardists, who would have wanted to make use of everything their trees produced. Fortunately, the tannic green walnuts made an excellent black dye, wood stain, and ink—but a liqueur made from the inedible fruit would have been valuable as well.

----

20 green walnuts, cut into quarters
1 cup sugar
750 ml bottle of vodka or Everclear
Zest of 1 lemon or orange
Optional spices: 1 cinnamon stick, 1 to 2 whole cloves,
    1 vanilla bean

*Green walnuts can be gathered in summer or purchased at farmers'
markets. Choose whole, unblemished fruit that can easily be pierced
with a knife. Wash them thoroughly before cutting. Pour the sugar in
a saucepan with just enough water to cover it and boil, stirring well.
Once the sugar is dissolved, combine it with the remaining ingredients in a large, sterile jar, and seal. Store it for 45 days in cool, dark
place, shaking occasionally. At the end of 45 days, strain out the
walnuts and spices and rebottle in a clean jar for another 2 months
of aging.*

*Some people add another cup of simple syrup before the last 2
months of aging. If you'd like to experiment, try splitting the batch
in two and adding ½ cup of simple syrup to one of those batches.
Either way, the aging process will change the flavor.*

The recipe for *nocino* (called *liqueur de noix* in France) has changed little over the centuries. It's nothing more than soft green walnuts, cut into quarters or crushed, and soaked in some kind of spirit along with sugar. Vanilla and spices can be added; some people include lemon or orange zest. It's ready to drink after it has sat for a month or two and turned a deep, rich brownish black.

Nocino doesn't have to be made at home. Haus Alpenz imports Nux Alpina Walnut Liqueur from Austria, and California's Charbay distillery makes a black walnut liqueur with a Pinot Noir brandy base called Nostalgie. Another California brandy-based walnut liqueur, Napa Valley's Nocino della Cristina, has also won rave reviews. Although *nocino* is intended to be sipped on its own as an after-dinner drink or poured over ice cream, bartenders also serve it in coffee drinks or in cocktails that call for spicy, nutty liqueurs.

PART III

At Last We Venture into the Garden,
Where We Encounter
a Seasonal Array of Botanical Mixers
and Garnishes to Be Introduced to the Cocktail
in Its Final Stage of Preparation

GARDENERS ARE THE ULTIMATE MIXOLOGISTS.

Even the most ordinary vegetable patch yields the mixers and garnishes that make remarkable drinks: it is nothing for a gardener to produce lemon verbena, rose geranium blossoms, sweet yellow tomatoes, and deep red stalks of heirloom celery. A thousand cocktails can be mixed in a kitchen garden.

Some plants, like mint for mojitos, should unquestionably be garden-grown. Others, like pomegranate for homemade grenadine, are worthwhile only if you live in a tropical climate or own a greenhouse and possess enough interest in horticulture to keep it alive and tend to its needs.

Rather than give the complete history, life cycle, and growing instructions for every plant that can be used as a mixer or a garnish, a few are highlighted in depth and others are simply listed along with a few growing tips. The best gardening advice is local anyway; whether a particular plant suits your climate or your level of expertise and commitment is a matter to be discussed with your local independent garden center, where you'll receive in-depth advice on planting varieties best suited to your area.

For even more information, turn to your local Master Gardener group (usually run through a county agricultural extension office) or the knowledgeable farmers at your farmers' market. Visit DrunkenBotanist.com for mail-order sources, growing tips, and further reading on culinary gardening.

-- we begin with --

HERBS

THESE HERBS CAN BE
muddled into a cocktail,
INFUSED INTO A SIMPLE SYRUP
OR FLAVORED VODKA,
AND USED AS A GARNISH.

Annual herbs live for just one year and require summer warmth, sun, and regular water, while woody, perennial herbs need sun and summer heat to thrive, prefer their soil on the dry side, and generally won't survive the winter if temperatures dip below 5 to 10 degrees Fahrenheit. Dedicated cold-climate gardeners keep perennial herbs in a pot and store them in the basement in winter with minimal light and water.

All of these herbs can live in a container, and most will grow indoors under bright lights. A conservatory or sunroom is ideal; even in a sunny window, they might need supplemental light indoors. An ordinary shop light with fluorescent tubes, plugged into a timer, is the most affordable solution. Garden centers and hydroponic shops also sell special grow lights and LED bulbs that screw into ordinary lamps, which may be somewhat more aesthetically pleasing.

The best way to harvest herbs is to cut one entire stalk down to the base of the plant, then strip the leaves from the stem. If you don't need that much, cut half the stalk. Just don't pluck off individual leaves: the plant can't easily regrow leaves from a bare stalk. Annual herbs tend to stop growing once they've flowered, so pinch off flowers on basil, cilantro, and other herbs you want to keep harvesting.

Angelica *Angelica archangelica*	Biennial (blooms in second year). Use stems in infusions. Other species may be toxic; be sure to get *A. archangelica*, called garden angelica. (See p. 140.)
Anise hyssop *Agastache foeniculum*	Perennial. Cut flowering stalks back to encourage rebloom. Try the bright yellow Golden Jubilee or the classic Blue Fortune. (See p. 180.)
Basil *Ocimum basilicum*	Annual. Genovese is the classic large-leaved variety. Pesto Perpetuo and Finissimo Verde are small-leaved, bush varieties that may overwinter indoors.
Cilantro *Coriandrum sativum*	Annual. Slow Bolt or Santo won't bloom and go to seed as quickly as other varieties. If you're growing it for coriander seeds, not cilantro leaves, look for *C. sativum* var. *microcarpum*. Seeds should be thoroughly dry and golden brown before using. (See p. 156.)
Dill *Anethum graveolens*	Annual. Dukat produces more foliage before setting seed. Fernleaf is a dwarf variety.
Fennel *Foeniculum vulgare*	Perennial. Both Florence and sweet fennel produce tasty seeds. Perfection and Zefa Fino are grown for their bulbs. (See p. 180.)
Lemongrass *Cymbopogon citratus*	Perennial. West Indian varieties are grown more for their stalks, East Indian for their leaves. Either works in cocktails.
Lemon verbena *Aloysia citrodora*	Perennial. A woody shrub that can grow four to five feet tall. Leaves have a bright, powerful citrus flavor. (See p. 175.)

Mint *Mentha spicata*	Perennial. Look for spearmints like *Mentha × villosa* 'Mojito Mint' or Kentucky Colonel. Other mints to experiment with include chocolate mint, orange mint, and peppermint. (See p. 325.)
Pineapple sage *Salvia elegans*	Perennial. A sturdy sage with red, trumpet-shaped flowers and leaves that really do smell of pineapples.
Rosemary *Rosmarinus officinalis*	Perennial. Arp is the most cold-resistant upright variety. Roman Beauty has higher oil content and a more compact habit. Avoid the prostrate or climbing varieties; the flavor is unpleasantly mentholated.
Sage *Salvia officinalis*	Perennial. Holt's Mammoth is a classic cooking variety. Any silver-leaved variety will work; the purple and yellow cultivars are not as flavorful.
Savory *Satureja montana*	Perennial. This is the winter savory, a woodier herb with a flavor closer to rosemary. The summer savory, *S. hortensis*, is used more as a fresh seasoning in eggs and salads.
Scented geranium *Pelargonium* sp.	Perennial. Although commonly called geraniums, they are actually pelargoniums. Breeders have created astonishing fragrances, from rose to cinnamon to apricot to ginger. Leaves are fragrant but strongly flavored; use for simple syrups and infusions. Blossoms make good garnishes.
Thyme *Thymus vulgaris*	Perennial. English thyme is the standard culinary thyme, but lemon varieties are also excellent. Creeping, woolly varieties are not as tasty.

GARDEN-INFUSED SIMPLE SYRUP

Almost any botanical ingredient, from lemon peel to rhubarb to rosemary, can be infused into a simple syrup. This is an easy way to showcase seasonal produce and add a twist to a basic cocktail recipe.

½ cup herbs, flowers, fruit, or spices
1 cup water
1 cup sugar
1 ounce vodka (optional)

Combine all the ingredients except the vodka in a saucepan. Bring to a simmer and stir well, until the sugar is dissolved. Let the mixture cool, then pour through a fine mesh strainer. Add the vodka (if using) as a preservative and keep refrigerated. Good for 2 to 3 weeks; lasts longer in the freezer.

SPANK YOUR HERBS

The secret to getting the essential oil out of any plant in the mint family (including mint, basil, sage, and anise hyssop) is to bruise the leaves without crushing them. This expresses the oil from the modified trichomes, or tiny hairs, on the surface of the leaf without cluttering up the drink unnecessarily with chlorophyll. Get the most flavor out of the fresh leaves by spanking them—just place the leaf in the palm of one hand and clap your hands briskly once or twice. You'll look like a pro and you'll release fresh aromatics into the drink.

SPEARMINT

Mentha spicata
LAMIACEAE (MINT FAMILY)

Thanks to the heroic efforts of tourists returning from Cuba with sprigs of mint plucked from their mojitos, mail-order nurseries now offer *Mentha × villosa* 'Mojito Mint' for sale, which they claim is distinctly different from most spearmints. "In a perhaps typically Cuban understated way its warm embrace lingers until you realize you want more," reads the catalog copy for this herb.

Never order a mojito in a bar that does not have fresh mint in evidence. Mint is so easy to grow that it is practically a weed; there is no excuse not to have a ready supply of it. Mint can live in a pot in the parking lot; it can grow in a window box; it can, for that matter, sprout in a rain gutter or between the cracks in the sidewalk.

Mint really will take over the garden if given a chance. To slow it down, plant it in a one-gallon plastic pot and sink the pot in the ground. The runners will find a way around the pot eventually, but at least you'll have a head start. Give the plant plenty of water—a perpetually damp spot near a leaky hose bib is perfect—and cut it before it blooms or goes to seed as the offspring tend to revert to a parent strain and won't be nearly as good. The flavor may change as the plant ages, so some gardeners root a runner every few years to replace an older plant.

The mint to grow is spearmint; it has a brightness and sweetness to it that seems to melt into sugar and rum. Look for 'Mojito Mint' or 'Kentucky Colonel,' the variety most favored by Southerners for a mint julep.

Spearmint, also called green mint, comes from central and southern Europe, where it has been cultivated for centuries. Pliny the Elder said that its scent "does stir up the mind." It also stirs up any number of drinks, adding a green and almost floral note to sweet and fruity cocktails that would otherwise be too cloying.

There are those who believe that a well-constructed mint julep is intended to last all day, that there is no second mint julep, just one large, powerful drink that grows gradually sweeter and more watered down as the ice melts and the sugar and bourbon settle together at the bottom of the glass.

Southern writer Walker Percy insisted that a good julep should hold at least 5 ounces of bourbon, a quantity that sloshes right over anyone's daily limit. This recipe remains true to his vision, but you may use less bourbon if you wish to feel like more of an upstanding citizen.

5 ounces bourbon
Several sprigs fresh spearmint
4 to 5 tablespoons superfine sugar
Crushed ice

Into a silver julep cup, a highball glass, or a Mason jar, press 2 or 3 tablespoons of superfine sugar together with a very small quantity of water, just enough to make a sugary paste. Add a layer of fresh spearmint leaves. Press them gently with a muddler or wooden spoon, but do not smash them. Pile on a layer of fresh finely crushed ice. Mr. Percy prefers that you reduce the ice to powder by wrapping it in a dry towel and banging it with a wooden mallet. To that layer, add a fine sprinkling of sugar and a few more mint leaves that you have spanked, but not crushed, by clapping them loudly between your hands.

Top with another layer of crushed ice and continue in this manner until the glass is so full that it seems that it cannot possibly hold a drop of bourbon. Pour in as much as it will, in fact, hold, which turns out to be right about 5 ounces. Now carry your julep to the porch and remain there until bedtime; there will be nothing else to your day but the slow draining of the glass and the pleasant drone of the cicadas.

-- moving on to --
FLOWERS

FLOWERS ARE MOST OFTEN USED
AS A GARNISH OR FROZEN IN ICE CUBES
FOR DECORATION, BUT SOME CAN BE
ADDED TO SIMPLE SYRUP OR
VODKA INFUSIONS TO *add flavor
or color.* THE BLOSSOMS OF THE
PLANTS IN THE HERBS SECTION (P. 320)
ARE ALSO EDIBLE AND SAFE TO USE.
JUST DON'T ADD ANY FLOWER
TO A COCKTAIL UNLESS YOU'RE SURE
IT'S EDIBLE. HYDRANGEA, FOR INSTANCE,
CONTAINS A LITTLE CYANIDE,
WHICH MAKES IT A LESS-THAN-IDEAL
DRINK INGREDIENT.

Borage *Borago officinalis*	Annual. Deep blue flowers are gorgeous in drinks or frozen into ice cubes. Leaves taste vaguely like cucumbers. Traditional Pimm's Cup garnish.
Calendula *Calendula officinalis*	Annual. Bright yellow and orange petals can be infused for their color. Alpha is a reliable orange variety; Sunshine Flashback is deep yellow; Neon is orange-red.
Elderflower *Sambucus nigra*	Perennial. Grown for flowers or fruit; use flowers in infusions and syrups. Try the dramatic Black Lace or Sutherland Gold, with chartreuse foliage. Some North American species produce cyanide, so get yours from a fruit nursery. (See p. 206.)
Honeysuckle *Lonicera* × *heckrottii*	Perennial. Gold Flame is tough, vigorous, and loaded with fragrant flowers.
Jasmine *Jasminum officinale*	Perennial. Hardy to about 0 degrees Fahrenheit. French jasmine, sometimes called *J. grandiflorum*, requires a warmer climate but can be grown indoors. (See p. 218.)
Lavender *Lavandula angustifolia*	Perennial. English lavenders like Hidcote and Munstead are best for culinary uses, or try the French Grosso and Fred Boutin (*L.* × *intermedia*). (See p. 330.)
Marigold *Tagetes erecta*	Annual. Petals are bright orange, red, or yellow and have a sharp, spicy flavor. There are many new varieties, but African Marigold is the classic vigorous orange version.

Nasturtium *Tropaeolum majus*	Annual. Dwarf Cherry is a mounding variety, compact enough for a container. Other varieties can turn into sprawling vines. All produce peppery flowers in orange, red, yellow, pink, and white.
Rose *Rosa* spp.	Perennial. Choose a highly fragrant hybrid tea like Mister Lincoln for rose petal infusions, or a rugosa variety if you want to cultivate the rose hips. (See p. 221.)
Sichuan button *Acmella oleracea*	Annual. The yellow flower buds contain a compound called spilanthol that produces a Pop Rocks–like reaction when chewed. A bit of a gimmick, but a fun cocktail garnish nonetheless.
Viola *Viola tricolor*	Annual. Johnny-jump-ups, and closely related pansies, are edible but not particularly flavorful. Useful as a garnish.
Violet *Viola odorata*	Perennial. Old-fashioned sweet violets are highly fragrant and very short-lived. Not to be confused with African violets. (See p. 225.)

☙ LAVENDER ❧

Lavandula angustifolia (syn. *L.* × *intermedia*)

LAMIACEAE (MINT FAMILY)

L avender doesn't often turn up behind the bar for the same reason it isn't used often in cooking: the sharp, floral fragrance seems just right in a perfume but all wrong as part of a meal. However, any gardener who loves growing lavender will eventually want to try it in a drink. And it is used in gin, infused vodka, and liqueurs.

English lavender, *Lavandula angustifolia,* is somewhat sweeter and better suited for flavoring—lavender scones and cookies are baked with this variety. Hidcote and Munstead are two popular varieties; both will grow up to two feet tall and form a solid hedge.

The only other lavender to consider for a cocktail would be *L.* × *intermedia,* a hybrid grown in France for perfume and soap. Try Grosso, Fred Boutin, or Abrialii. The flavor may be a bit sharper than English lavender, but the plants do a better job of tolerating hot, muggy summers. Many other species of lavender contain mildly toxic compounds and should not be eaten.

LAVENDER-ELDERFLOWER CHAMPAGNE COCKTAIL

1 ounce lavender simple syrup (see p. 324)
1 ounce St-Germain
Champagne or other sparkling wine
1 sprig fresh lavender

Pour the simple syrup and St-Germain in a Champagne flute and top with Champagne. Garnish with a sprig of fresh lavender.

Situate lavender in full sun, give it well-drained soil, and top-dress with pea gravel rather than mulch. It wants no fertilizer and very little supplemental water. Lavender must be sheared back in late fall to keep blooming; cut back most of the foliage, but never cut down to bare wood. Lavender likes a Mediterranean climate but can be coaxed along in all but the coldest areas; it will tolerate winter temperatures to −10 degrees Fahrenheit.

Lavender's dry, astringent perfume is perfect with a botanical spirit like gin, or it can be used to infuse a simple syrup.

LAVENDER MARTINI

4 sprigs fresh lavender
1½ ounces gin (try Dry Fly from Washington State,
    which contains lavender)
½ ounce Lillet blanc (see note)
Lemon peel

In a cocktail shaker, muddle 3 springs of the lavender with the gin. Add the Lillet, shake with ice, and strain into a cocktail glass. To get all the crushed lavender buds out of the drink, double-strain it by placing a fine mesh strainer over the glass before pouring. Garnish with the lemon peel and the remaining lavender sprig.

NOTE: LILLET WILL STAY FRESH IN THE REFRIGERATOR FOR AT LEAST A FEW WEEKS. IF YOU DON'T HAVE ANY, THIS DRINK ALSO WORKS WITH THE MORE TRADITIONAL DRY VERMOUTH.

-- continuing straight ahead to --
TREES

FRUIT TREES ARE NOT EXACTLY
AN IMPULSE PURCHASE.
a tree is like a puppy:
IT'S CUTE WHEN IT'S SMALL,
BUT IT DOES GROW UP,
AND IT REQUIRES A LIFETIME OF CARE.

Some fruit trees need a certain number of winter chill hours (hours during which temperatures hover around freezing) in order to complete their dormancy cycle. Some are vulnerable to pests and diseases that would require you to apply more sprays than you might be comfortable with, considering the fact that cocktails often call for the rind of the fruit. Consult with a good local fruit tree nursery or your county agricultural extension office, both of whom might offer fruit tree workshops, and ask about disease and pest-resistant varieties and organic methods.

Some trees, including citrus, will grow in a container and won't object to overwintering indoors if they can't survive the winter outside. Just know that fruit trees tend to be grafted onto rootstock that determines their size, so if you want a tree to stay small, ask for one grafted onto dwarf rootstock.

The care and feeding of a fruit tree is also a little different from other plants. Some varieties are self-fertile, which means they don't need a mate nearby, but others won't bloom without a compatible tree (called a pollinizer) in the area. Speaking of pollination, your local bees will probably do the deed without any special effort on your part, but indoor plants might require a little assistance. (Have this birds-and-bees talk with your garden center staff.) Fruit trees also require special fertilizer loaded with micronutrients like iron, copper, and boron. They call for particular pruning strategies, and some must be thinned when the fruit is still small and green to ensure a good harvest.

None of this, however, should discourage you. Fruit trees are endlessly rewarding. Some nurseries will double or triple that reward by grafting a few varieties onto one rootstock. These "three in one" or "four in one" offerings are a great way to grow a variety of fruit in a small space. With a little education and help selecting the right variety for your area, you'll enjoy the singular pleasure of enlivening a drink with fresh, seasonal juice from your own garden.

Apple *Malus domestica*	Choosing varieties that thrive in your climate is key. Do plenty of tasting at farmers' markets and ask local growers for help selecting a tree. (See p. 17.)
Apricot *Prunus armeniaca*	Most apricots grown for fruit in the United States have bitter, inedible pits, which is fine if you only intend to use the fruit. One sweet variety, whose almond-flavored pits can be soaked in brandy, is SweetHeart. (See p. 260.)
Cherry *Prunus cerasus* var. *marasca*	To make your own maraschino cherries, look for a sour, dark, morello type, also called a pie cherry. (See p. 271.)
Fig *Ficus carica*	Violette de Bordeaux is a classic French variety, but what matters most is choosing one suitable for your area. Try to taste figs from local farmers before committing to a variety. Great for simple syrup reductions. (See p. 270.)
Lemon *Citrus limon*	Great in containers. Choose Improved Meyer for the juice or Eureka or Lisbon for flavorful rind. (See p. 297.)
Lime *Citrus aurantifolia*	Also called key lime, Mexican lime, or West Indian lime, this is the ideal variety for mixed drinks. Kaffir lime, *C. hystrix*, is grown for its aromatic leaves, which are used in Thai-flavored drinks. (See p. 298.)
Lychee *Litchi chinensis*	An extraordinary tropical fruit, the juice is delightful in cocktails and the fruit makes a lovely garnish. The tree can't survive temperatures below 25 degrees Fahrenheit and grows to over thirty feet, making it unsuitable for cold climates or greenhouses.

Olive *Olea europaea*	Gordal is a classic Spanish variety. Arbequina is small and cold-tolerant. Look for a variety that is bred for fruit production, not as an ornamental tree. Be aware that olive pollen can be highly aggravating to people with seasonal allergies. (See p. 336.)
Orange *Citrus aurantium,* others	The so-called bitter orange is grown for its peel, as are citrons. Navel and blood oranges are better for juice, and some varieties will grow indoors. For containers, consider kumquats and calamondins for sheer ease of growing. (See pp. 287, 301.)
Peach *Prunus persica*	Look for disease-resistant, dwarf varieties. Peaches (and closely related nectarines) are ideal for so-called combo trees, where a few varieties are grafted to one rootstock.
Plum *Prunus domestica*	Dark blue damson, bright yellow mirabelle, and greengage plums are the traditional European varieties used for making wines, liqueurs, and eau-de-vie. Try Big Mackey or Jam Session, bred by Cornell University to succeed in North America. (See p. 277.)
Pomegranate *Punica granatum*	The dwarf variety *P. granatum* var. *nana* thrives in pots, but commercial growers prefer Wonderful, the variety grown by the founders of the juice company POM Wonderful, which also supplies fresh pomegranates to markets around the world. Angel Red and Grenada ripen earlier than Wonderful, making them more likely to set fruit before an early freeze arrives. (See p. 337.)

BRINE YOUR OWN OLIVES

A terrible olive can ruin a great martini. If you have access to fresh olives, try brining your own with nothing but water and salt.

Get fresh-picked green olives from a farmers' market (or a friend with an olive tree) and make a single cut in each one from top to bottom. Wash them in plain water and place them in a clean glass jar or bowl. Choose your container carefully; you'll need to weigh the olives down, so pick something with a wide mouth and find a plate or lid that just fits inside. (A sturdy plastic bag filled with water also works as a weight.) Soak the olives in water for 24 hours, being sure to keep them submerged. Keep them stored in a cool, dry place during this process.

Change the water every day for 6 days. After 6 days, make the final brine by combining 1 part pickling salt to 10 parts water in a saucepan. Bring to a boil and allow to cool. Pour the olives into jars and fill with brine. Add lemon, garlic, spices, or herbs, if desired. Seal them tightly and refrigerate for 4 more days. They'll be ready to eat and should be kept refrigerated and enjoyed fresh.

POMEGRANATE

Punica granatum

LYTHRACEAE (LOOSESTRIFE FAMILY)

An 1867 medical journal entry on pomegranate explained that "the tincture, a liqueur glass morning and evening, infallibly expels the yellow tapeworm." This was not the first report of the vermifugal powers of pomegranate: a Portuguese doctor had been making a tea of the bark for the same purpose since 1820 and calling it grenadine. Fortunately, by the second half of the nineteenth century, grenadine had come to refer to a sweet, ruby-colored fruit syrup used to flavor sodas and drinks, not a bark tea designed to kill intestinal worms.

A pomegranate tree is actually a large shrub of Asian and Middle Eastern origin. It still grows wild there today, but now it is cultivated throughout Europe, the Americas, and in tropical areas around the world. Although the tree has ancient origins and was extensively grown by Egyptians, there are just two species. The plants were once classified in a taxonomic family by themselves, until new molecular research uncovered their close genetic relationship to purple loosestrife, crepe myrtle, cuphea, and other seemingly dissimilar plants. (Crumpled flower petals are their most obvious shared anatomical feature.)

Pomegranate trees are now grown primarily in the Middle East, India, and China, although they are a specialty crop in the Mediterranean, and in Mexico and California. The fruit earned its species name *granatum* from the Latin word for "seeded," and its fruit does, in fact, contain a few hundred seeds surrounded by bright red pulp. The syrup made from it, grenadine, derives from the early French word for pomegranate, *grenade*. The hand-thrown projectile of the same name was invented in the sixteenth century and was named after the fruit, perhaps because they were each the same size and filled with explosive materials of a very different kind.

HOMEMADE GRENADINE

5 to 6 fresh pomegranates
1 to 2 cups sugar
1 ounce vodka

To peel the pomegranates, score the rind with a knife as if you're cutting an orange into wedges. Carefully peel away the rind, leaving the seeds and membrane intact. Squeeze with a fruit press or manual juicer and filter through a sieve. You should have about 2 cups of juice.

Measure 1 cup of the sugar into a saucepan, add the juice, stir, and bring to a simmer. Let the sugar cool and taste it; add more sugar if you prefer a sweeter syrup. Stir in the vodka as a preservative. Pour into a clean jar and store it in the refrigerator, where it will last about a month, or in the freezer. Adding another ounce or two of vodka will help keep it from freezing.

Grenadine syrup was popular in French cafes as a sweetener for water in the 1880s, and shortly thereafter it began turning up in American soda fountains and cocktail bars. In 1910, the St. Regis hotel in New York served a cocktail called the Polly made of gin, grenadine, lemon juice, and soda. In 1913, the *New York Times* sent a skeptical male reporter to a bar designed just for women called Café des Beaux Arts on Sixth Avenue at Fortieth Street. Among the many marvels he discovered in this feminine establishment were brightly colored cocktails, including the frothy pink Beaux Arts Fizz, made with gin, orgeat (sweet almond) syrup, grenadine, and lemon juice.

Grenadine's run as a pure pomegranate syrup was surprisingly short. Artificial versions appeared in the early twentieth century, and by 1918 manufacturers were challenging new labeling laws, trying to pass off any sort of red syrup as grenadine. As one reporter described the situation, "The syrup and the fruit from which it took its name were total strangers." Although the artificial form eventually won out, grenadine stayed behind the bar, becoming an essential ingredient in hundreds of cocktails, including the Jack Rose and the tiki classic, Tequila Sunrise.

Thanks to a renewed interest in authentic ingredients, grenadine made with actual pomegranates can now be found on the shelves of better liquor stores and specialty food shops, as can pomegranate liqueurs and infused vodkas. But there is no substitute for homemade grenadine made from fresh-squeezed pomegranates. Even replacing the fresh juice with bottled compromises the flavor. When the fruit is in season, it is well worth spending an hour or so in the kitchen making up a batch for the freezer.

JACK ROSE

1½ ounces **applejack**
½ ounce **fresh lemon juice**
½ ounce **grenadine**

Shake all the ingredients over ice and strain into a cocktail glass.

-- proceeding onward to --
BERRIES & VINES

JUST ABOUT EVERYTHING THAT CAN
BE SAID OF FRUIT TREES
APPLIES TO BERRIES AND VINES
AS WELL, *with one exception:*
THESE PLANTS ARE RELUCTANT
TO GROW IN CONTAINERS
AND WON'T ENJOY LIFE INDOORS.

Berries tend to be low-maintenance, requiring only a trellis, once-a-year pruning, and occasional fertilizer. Most berries are planted in bare-root form (you buy a clump of live roots attached to a cane, not a growing plant) in winter or early spring.

Do check with local experts to find the varieties that grow best in your climate, and ask if you need a nearby pollinizer. Get pruning advice specific to the variety you've chosen: some raspberries, for instance, bear fruit twice a year and only on two year-old canes, which means that you have to cut back the old canes after they've fruited, but let the younger canes grow unmolested for two years before they bear fruit.

berries & vines

Blackberry *Rubus* spp.	Do yourself a favor and choose a thornless variety. Extend the growing season by selecting several cultivars with different blooming periods; for instance, Arapaho starts in mid-June, and Black Diamond bears in August. Loganberry, marionberry, boysenberry, and tayberry crosses (generally blackberry-raspberry hybrids) are well worth growing.
Blueberry *Vaccinium* spp.	Because blueberries prefer acid, moist soil, growing them in a container might be the best way to give them the conditions they need. Top Hat and Chippewa are compact varieties for pots. Some varieties tolerate winter temperatures to −20 degrees Fahrenheit.
Currant *Ribes nigrum*	The black currant, used to make cassis, is still banned in some states even though new disease-resistant cultivars do not spread the dreaded white pine blister rust. Ben Lomond is a vigorous Scottish variety. Red and white currants have a bright, light flavor and make beautiful garnishes in drinks. (See p. 263.)
Hops *Humulus lupulus*	Hops require specific day lengths to bloom, so do best in latitudes of 35 to 55 degrees north and south. The golden hop vine Aureus, with its yellow to lime green foliage, is a widely sold ornamental, as is Bianca, a variety with light green foliage that matures to a darker green. (See p. 210.)
Raspberry *Rubus idaeus*	Look for everbearing varieties that fruit over a long season. Pruning is simpler because all canes are cut down each winter. Try Caroline or Polka Red.
Sloe *Prunus spinosa*	Also called blackthorn, this large, thorny shrub is hardy to −30F. Bears the fruit used to make sloe gin— if the birds don't get them first. (See p. 281.)

INFUSED VODKAS

Nothing could be simpler than infusing herbs, spices, and fruit in vodka to make your own flavored spirit for cocktails. There's just one catch: some plants, particularly tender green herbs like basil or cilantro, produce bitter, strange flavors if they've been soaking for long. To get around this, make a small batch as a test, and taste it frequently, starting just a few hours after the infusion has begun. For herbs, 8 to 12 hours may be sufficient. For fruit, 1 week may be long enough. Citrus rinds and spices can soak for a month. The trick is that as soon as it tastes wonderful, strain it. Time will not necessarily improve an infusion.

The instructions are simply this:

Fill a clean jar with herbs, spices, or fruit. Pour in an affordable but not rock-bottom vodka, like Smirnoff. Seal tightly and store in a cool, dark place. Sample regularly until you decide it's perfect. Strain it and use it within a few months.

LIMONCELLO AND OTHER LIQUEURS

Consider this recipe to be a template for other sweet infusions. Coffee beans, cocoa nibs, or almost any kind of citrus could take the place of lemon to make another sweet, after-dinner liqueur.

12 fresh lemons (see note)
1 750 ml bottle vodka
3 cups sugar
3 cups water

Peel the lemons, being careful to get only the yellow rind. (If you don't have another use for the fruit, squeeze the juice and freeze in ice cube trays for use in cocktails.) Place the lemon rinds and vodka in a large glass pitcher or jar. Cover and let sit for 1 week.

After 1 week, heat the sugar and water, let it cool, and add it to the vodka and lemon mixture. Let it sit for 24 hours, and then strain. Refrigerate overnight before drinking.

NOTE: AVOID CHEMICALS AND SYNTHETIC WAXES BY CHOOSING ORGANIC OR UNSPRAYED, HOMEGROWN CITRUS.

-- and wrapping things up with --
FRUITS & VEGETABLES

ANY WELL-STOCKED KITCHEN GARDEN
COULD EASILY SUPPLY A BARTENDER
AS WELL, BUT IF YOU'RE FOCUSING
EXCLUSIVELY ON DRINKS, YOU CAN
FORGO CULINARY NECESSITIES LIKE
SALAD GREENS AND SUMMER SQUASH,
AND INSTEAD PLANT A *fruit and
vegetable garden made
exclusively for cocktails.*
LOOK FOR VARIETIES THAT PRODUCE
OVER A LONG GROWING SEASON,
OR FIND EARLY- AND LATE-SEASON
VARIETIES OF THE SAME FRUIT OR
VEGETABLE, TO EXTEND THE HARVEST.
LOOK FOR SMALL-FRUITED VARIETIES,
TOO. AFTER ALL, MOST DRINKS CALL FOR
ONLY SMALL QUANTITIES OF FRUIT, AND
COCKTAIL GLASSES THEMSELVES CAN
ONLY ACCOMMODATE A DIMINUTIVE
GARNISH BEFORE THEY GET DIFFICULT TO
DRINK. HERE ARE A FEW FAVORITES.

fruits & vegetables

Celery *Apium graveolens*	Believe it or not, celery is well worth growing if you have a long, cool growing season. Homegrown stalks may be thinner than beefy, store-bought varieties, which make them perfect as swizzle sticks. Look for the dramatic crimson Redventure.
Cucumber *Cucumis sativus*	Spacemaster 80 and Iznik do well in containers, Corinto tolerates heat waves or unexpected cold spells; Sweet Success resists diseases and is a 'burpless' variety, also called an English variety, which is said to be easier to digest.
Melon *Cucumis melo*	The best way to select a variety to plant might be to buy a selection from the farmers' market and save the seeds of your favorite. Ambrosia resists powdery mildew; Charentais is a classic French variety.
Miracle fruit *Synsepalum dulcificum*	A good container plant, available from tropical plant nurseries. This native West African shrub produces tiny dark red berries that contain a glycoprotein with a strange effect on the tongue: When eaten, the proteins bind to taste buds and change the way receptors perceive flavor. For about an hour, until digestive enzymes break down the proteins, sour foods taste sweet. The possibilities have not been lost on bartenders; a sour drink made with lemon or lime juice can be garnished with a miracle berry, the idea being that bar patrons will take a few sips, eat the berry, and enjoy a completely different cocktail afterward. Fresh berries are hard to find unless you grow them yourself.
Pepper *Capsicum annuum*	Easy to grow in containers, requiring only heat and light. Try sweet varieties like Cherry Pick and Pimento-L and hot varieties like Cherry Bomb and the Peguis jalapeño. Great for garnishes and infused vodkas. (See p. 352.)

Pineapple *Ananus comosus*	Pineapples grow out of the center of a small, bromeliad-like plant. Best grown in a container indoors, they take two years to produce fruit. Royale is a smaller variety better suited to home gardeners.
Rhubarb *Rheum rhabarbarum*	Rhubarb simple syrup is a must-have cocktail ingredient. Give it a permanent spot with rich, loamy soil and it will produce for years. Eat the stalks only; the leaves are poisonous.
Strawberry *Fragaria × ananassa*	Perfect container plants; strawberries thrive in hanging baskets or strawberry pots with regular water. Strawberries grown in the ground should be mulched with straw to protect the fruit from coming in contact with soil and rotting. Look for everbearing or day neutral varieties that produce over a long season. Tiny alpine strawberries (*F. vesca*) are small, tart varieties that make beautiful garnishes and also bear over a long season. (See p. 350.)
Tomatillo *Physalis philadelphica*	These tart green fruits are essential to salsa verde but are fantastic muddled into tequila cocktails as well. Toma Verde is a classic green variety; Pineapple is bright yellow with a tropical, pineapple flavor.
Tomato *Solanum lycopersicum*	Juicy, ripe tomatoes pair perfectly with vodka and tequila. Sungold is everyone's favorite cherry tomato, and Yellow Pear also makes a beautiful garnish. New grafted varieties are grown on vigorous, disease-resistant rootstock; although the plants cost more, they may be tougher and offer better yields.
Watermelon *Citrullus lanatus*	Watermelon is divine in rum, tequila, or vodka drinks. Faerie is a yellow-skinned, red-fleshed variety that produces small fruit and resists disease. Little Baby Flower is another disease-resistant variety that produces lots of small fruit rather than a few enormous melons.

GARDEN COCKTAILS:
A TEMPLATE *for* EXPERIMENTATION

With a garden full of fresh produce, you hardly need a recipe to mix an amazing cocktail. Just use a few basic proportions to combine ingredients and make a balanced drink. Here are a few examples to get you started:

1½ OUNCES OF	MUDDLED WITH	PLUS A BIT OF	AND IF YOU WANT TO GET FANCY, A SPLASH OF	THEN SERVE IT
Bourbon	Peaches and mint	Simple syrup	Peach bitters	In a Mason jar over crushed ice
Gin	Cucumber and thyme	Lemon juice	St-Germain	Shaken and served over ice with tonic
Rum	Strawberry and mint	Lime juice and simple syrup	Velvet Falernum	Over ice, topped with soda or sparkling wine
Tequila	Watermelon and basil	Lime juice	Cointreau	Shaken in a cocktail glass
Vodka	Tomato and cilantro	Lime juice	Celery bitters	Shaken in a cocktail glass

REFRIGERATOR PICKLES

Cucumbers, green beans, asparagus, carrots, Brussels sprouts, celery, green tomatoes, zucchini, pearl onions, yellow beets, and okra all make fine cocktail garnishes. This quick pickle recipe requires no special equipment—just remember that the pickles must be refrigerated and will only last 2 to 3 weeks.

2 cups sliced or cubed vegetables
2 teaspoons coarse, noniodized salt
2 cups sugar
1 cup cider or white vinegar
1 teaspoon each pickling spices (such as seeds of dill, celery, mustard, fennel)
Lemon rind, onion slices, garlic slices (optional)

Slice or cube the vegetables according to the kind of garnish you'd like to make. Toss with the salt and set aside for 30 to 45 minutes. Heat the sugar and vinegar in a saucepan until the sugar dissolves; let cool.

Fill clean jars with the vegetables, pickling spices, and optional ingredients, if desired. Top off the jars with the vinegar mixture, seal tightly, and refrigerate overnight.

✦ STRAWBERRY ✦

Fragaria × ananassa
ROSACEAE (ROSE FAMILY)

T hose large, juicy red strawberries in your summer cocktail owe
their unlikely existence to a French spy, a global voyage, and a
serious case of gender confusion.

In 1712, an engineer named Amédée François
Frézier was sent to Peru and Chile to make
a reliable map of the coast for the French
government. The area was under Spanish control, so to get the information
he needed, he posed as a traveling
merchant. He made a number of
useful maps, but he also did a little
botanizing while he was there.
Although tiny native wild strawberries (including *Fragaria vesca,* the
alpine strawberry, and *F. moschata,*
the highly flavorful musk strawberry)
were already cultivated in Europe, no one
had ever seen a strawberry as large as the
Chilean species, *Fragaria chiloensis.*

He collected as many plants as he could, but only five survived the
voyage home. Two went to the ship's cargo master as an expression of thanks for letting him use some of the ship's limited supply
of fresh water to care for the plants. One went to his supervisor
and one to the Jardin des Plantes in Paris, leaving him with a single
plant.

European botanists were happy to have the Chilean strawberry,
but there was one problem: it was sterile. The only way to get more
plants was to divide it. What Frézier didn't know is that Chilean
strawberries can be male, female, or bisexual. He chose the plants

that were bearing the largest fruit, and those happened to all be female. They needed males nearby to reproduce and create larger, more luscious fruit.

Eventually farmers realized that the males of other strawberry species could do the job. By the mid-nineteenth century, the Chilean strawberry had been crossed with a native Virginian species that has also been brought to Europe, *F. virginiana,* and the modern strawberry was born.

THE FRÉZIER AFFAIR

This daiquiri variation uses Chartreuse as a nod to Amédée François Frézier's French heritage. The yellow version is sweeter, but if you only have the green version, use it along with a dash of simple syrup. St-Germain, an elderflower liqueur, is another good substitute.

3 slices ripe strawberry
1½ ounces white rum
½ ounce yellow Chartreuse
Juice of 1 fresh lemon wedge

Reserve 1 strawberry slice for garnish. In a cocktail shaker, combine the remaining ingredients and crush the strawberries with a muddler. Shake over ice, and strain into a cocktail glass. Garnish with the reserved strawberry slice.

PEPPER

Capsicum annuum, C. frutescens

SOLANACEAE (NIGHTSHADE FAMILY)

This tropical American plant was domesticated by native people fifty-five hundred years ago. A species called the wild bird pepper (*Capsicum annuum* var. *aviculare*) still grows in South and Central America and is believed to be the most similar to the original pepper in its wild, unadulterated state. The plant produces tiny fruit, each about the size of a raisin and shockingly hot.

The Aztecs called these peppers *chilli*. When Columbus arrived, thinking he'd reached India, he called the dried, shriveled fruits "peppers" because they resembled black pepper from India. Once the plants reached Europe, Spaniards attempted to rename them *pimento* to clear up the confusion. That name was applied to a particular kind of sweet pepper still popular in Spain, but otherwise the name pepper (or chili pepper) stuck.

THE PEPPADEW MYSTERY

PEPPADEW IS THE BRAND NAME of a kind of marinated sweet pepper that its manufacturer calls a sweet piquanté pepper. According to the company, a man named Johan Steenkamp discovered the plant growing in the backyard of his summer home in Tzaneen, South Africa. The jarred peppers have become so popular in cocktails and appetizers that gardeners have gone mad trying to find the seeds, but the company keeps the variety name a secret and has claimed international breeders' rights to control access to the pepper. Until Peppadew divulges its secrets, try growing Cherry Pick or, for a little more heat, Cherry Bomb.

A pepper is a fruit filled with air instead of juicy flesh. More specifically, it's a berry, a berry being a single ovary containing seeds, but only a botanist would ever call it that. The fruit gets its heat from capsaicin, a substance that appears in highest concentrations in the inner membrane of the fruit and in its seeds. While capsaicin doesn't cause a physical burn, it does send a signal to the brain that something's on fire. The brain responds by issuing pain signals in an attempt to persuade the body to get away from the fire—fast.

The brain also releases a flood of endorphins, or natural painkillers, when it believes an injury like a burn has occurred. For that reason, hot peppers can deliver a genuine sense of euphoria—even when they're not in a cocktail.

Peppers require rich soil, warm temperatures, bright sunlight, and regular water to flourish. Gardeners should choose their cocktail varieties according to taste; there's no reason to grow fresh jalapeños if you can't stand the heat.

CAYENNE: **A HOT SPICE MADE FROM CRUSHED, DRIED CAYENNE PEPPERS**

PAPRIKA: **A MILD SPICE MADE FROM CRUSHED, DRIED SWEET PEPPERS**

BLUSHING MARY

1½ ounces vodka or tequila
4 to 5 cherry tomatoes, halved
1 mild or hot pepper, sliced
2 dashes Worcestershire sauce
2 to 3 leaves of basil, parsley, cilantro, or dill
4 ounces tonic water
Celery bitters
Cracked black pepper (optional)
Slice of pepper, cherry tomato, herb leaf, celery stick,
    or olive for garnish

Combine the first five ingredients in a cocktail shaker; then use a muddler to crush the vegetables and herbs. Shake well over ice and strain into an Old-Fashioned glass filled with ice. Add the tonic water and stir. Finally, add a dash of celery bitters, crack a bit of black pepper on top, if desired, and add your choice of garnish. Vegetarians wishing to avoid the anchovies in regular Worcestershire can try the Annie's Naturals brand.

DIGESTIF

VINTNERS, BREWERS, DISTILLERS, AND BARTENDERS
are an endlessly inventive lot. The cocktail revival that is under way
in the first part of the twenty-first century, coupled with a renewed
interest in fresh, local ingredients, means that drinkers will be
treated to an ever-changing menu of interesting drinks. Obscure
plants will come into vogue, long-forgotten herbal ingredients will
be revived, and new, improved varieties will make it easier than ever
to grow a damson plum or a black currant in your own backyard.

The end of this book is only the beginning of a dialogue about botany and booze. Visit me at DrunkenBotanist.com for plant and liquor
sources, bibliographies and recommended reading lists, botanical
cocktail events, farm-to-distillery tours, recipes, and techniques for
both gardeners and mixologists. If you've got a question, a quarrel,
a good gin recommendation, or a new horticultural discovery of
your own, drop me a note through the website. I'd love to continue
the conversation over a good drink. *Salud!*

RECOMMENDED READING

recipes

Beattie, Scott, and Sara Remington. *Artisanal Cocktails: Drinks Inspired by the Seasons from the Bar at Cyrus*. Berkeley, CA: Ten Speed Press, 2008.

Craddock, Harry, and Peter Dorelli. *The Savoy Cocktail Book*. London: Pavilion, 1999.

DeGroff, Dale, and George Erml. *The Craft of the Cocktail: Everything You Need to Know to Be a Master Bartender, with 500 Recipes*. New York: Clarkson Potter, 2002.

Dominé, André, Armin Faber, and Martina Schlagenhaufer. *The Ultimate Guide to Spirits & Cocktails*. Königswinter, Germany: H. F. Ullmann, 2008.

Farrell, John Patrick. *Making Cordials and Liqueurs at Home*. New York: Harper & Row, 1974.

Haigh, Ted. *Vintage Spirits and Forgotten Cocktails: From the Alamagoozlum to the Zombie and Beyond: 100 Rediscovered Recipes and the Stories Behind Them*. Beverly, MA: Quarry Books, 2009.

Meehan, Jim. *The PDT Cocktail Book: The Complete Bartender's Guide from the Celebrated Speakeasy*. New York: Sterling Epicure, 2011.

Proulx, Annie, and Lew Nichols. *Cider: Making, Using & Enjoying Sweet & Hard Cider*. North Adams, MA: Storey, 2003.

Regan, Gary. *The Joy of Mixology*. New York: Clarkson Potter, 2003.

Thomas, Jerry. *How to Mix Drinks, or, The Bon Vivant's Companion: The Bartender's Guide*. London: Hesperus, 2009.

Vargas, Pattie, and Rich Gulling. *Making Wild Wines & Meads: 125 Unusual Recipes Using Herbs, Fruits, Flowers & More.* North Adams, MA: Storey, 1999.

Wondrich, David. *Imbibe! From Absinthe Cocktail to Whiskey Smash, a Salute in Stories and Drinks to "Professor" Jerry Thomas, Pioneer of the American Bar.* New York: Perigee, 2007.

gardening

Bartley, Jennifer R. *The Kitchen Gardener's Handbook.* Portland, OR: Timber Press, 2010.

Bowling, Barbara L. *The Berry Grower's Companion.* Portland, OR: Timber Press, 2008.

Eierman, Colby, and Mike Emanuel. *Fruit Trees in Small Spaces: Abundant Harvests from Your Own Backyard.* Portland, OR: Timber Press, 2012.

Fisher, Joe, and Dennis Fisher. *The Homebrewer's Garden: How to Easily Grow, Prepare, and Use Your Own Hops, Brewing Herbs, Malts.* North Adams, MA: Storey, 1998.

Hartung, Tammi. *Homegrown Herbs: A Complete Guide to Growing, Using, and Enjoying More Than 100 Herbs.* North Adams, MA: Storey, 2011.

Martin, Byron, and Laurelynn G. Martin. *Growing Tasty Tropical Plants in Any Home, Anywhere.* North Adams, MA: Storey, 2010.

Otto, Stella. *The Backyard Orchardist: A Complete Guide to Growing Fruit Trees in the Home Garden.* Maple City, MI: OttoGraphics, 1993.

Otto, Stella. *The Backyard Berry Book: A Hands-on Guide to Growing Berries, Brambles, and Vine Fruit in the Home Garden.* Maple City, MI: OttoGraphics, 1995.

Page, Martin. *Growing Citrus: The Essential Gardener's Guide.*
Portland, OR: Timber Press, 2008.

Reich, Lee, and Vicki Herzfeld Arlein. *Uncommon Fruits for Every Garden.* Portland, OR: Timber Press, 2008.

Soler, Ivette. *The Edible Front Yard: The Mow-Less, Grow-More Plan for a Beautiful, Bountiful Garden.* Portland, OR: Timber Press, 2011.

Tucker, Arthur O., Thomas DeBaggio, and Francesco DeBaggio. *The Encyclopedia of Herbs: A Comprehensive Reference to Herbs of Flavor and Fragrance.* Portland, OR: Timber Press, 2009.

ACKNOWLEDGMENTS

I OWE A ROUND OF DRINKS to the many distillers, bartenders, botanists, anthropologists, historians, and librarians who took the time to answer my questions, share their work, and help me chase down obscure facts. This is only a partial list, but here it goes: in the booze world, thanks to Alain Royer and his French connections, Bianca Shevlin at SABMiller, Don Poffenroth at Dry Fly Distilling, Mrs. Loes van der Woude at Curaçao, Melkon Khosrovian of Greenbar, Tyler Schramm of Pemberton Distillery, Tom Burkleaux at New Deal Distillery, Matt Mount at House Spirits, Eric Seed and Scott Krahn at Haus Alpenz, Joel Elder and Gable Erenzo at Tuthilltown, Isabella D'Anna at Verviene du Velay and the Cassissium, the legendary Stephen McCarthy of Clear Creek Distillery, the incomparable Jacqueline Patterson at Lillet, Allison Evanow at Square One, everyone at St. George Spirits, Jose Hermoso at International Wine & Spirit Research, David Williamson of the Scotch Whisky Association, Matt Colglazier of Sorgrhum fame, Woodford Reserve master distiller Chris Morris, Scott Goldman at Cadre Noir Imports, David Suro-Piñera at Sierra Azul, Greg Lorenz at SakeOne, Debbie Rizzo of DrinkPR, Nathan Greenawalt of Old Sugar Distillery, and Avery Glasser of Bittermens.

On the academic and botanical side, thanks go to: SUNY Buffalo law professor Mark Bartholomew for a fascinating discussion of trademarks; the USDA's David H. Gent for help on hops; Amy Iezzoni at Michigan State University for insight into cherries; Scott Calhoun, Greg Starr, and Randy Baldwin for cactus and agave expertise; Stark Bros. Nursery for damson plums; University of British Columbia's Michael Blake for his sugarcane research; Alan Fryar of University of Kentucky for his expertise in limestone; Stuart Swanson of the Scottish Crop Research Institute; hop farmers Darren Gamache and Gayle Goschie; archeologist Patrick McGovern; James Luby at University of Minnesota on grapes; Lena Struwe and Rocky Graziose at Rutgers on gentian; Jeff Gillman at University of Minnesota for botanical inquiries of all sorts; Cornell pomologists Ian Merwin and Susan Brown; Cornell's champion of the black currant Steven

McKay; Véronique Van de Ponseele at France's Muséum National d'Histoire Naturelle; Humboldt State chemistry professor Kjirsten Wayman; Tom Elias and Jacquelyn Kallunki at New York Botanical Garden on angostura; Laura Ackley for her Panama Exposition expertise; the German translation team at Filomel and the French translation team of Vic Stewart and Guy Vicente; Kandie Adkinson at Kentucky's Office of the Secretary of State; superstar librarians Axel Borg at UC Davis; Linda L. Oestry at Missouri Botanical Garden; the Bancroft Library staff; and Matthew Miles and everyone else at the Humboldt County Library and Humboldt State University library.

INDEX

α-crocin, 224
α-pinene, 171
α-terpineol, 298
Abbott's Bitters, 197
Abipón tribe, 117
Abrialii lavender, 330
absinthe, 40, 66, 106, 178, 200–3
Absolut vodka, 109
acacias, 252
Acer saccharum, 257–58
  palmatum, 258
achiote tree, 269
acholado, 69
Acmella oleracea, 329
acocote, 3
Acorus calamus (*A. americanus*), 148
Acremonium, 84
Adams, John, 101
Adiantum capillus-veneris, 185–87
Aframomum melegueta, 168
agaric, 233
Agastache foeniculum, 180, 322
agava, 15
agave, 2–16
  bugs in booze, 16
  field guide to, 11
  French intervention, 13
  list of spirits, 15
  mezcal, 4–10
  100%, 11
  protecting the plants, 10, 13–14
  pulque, 2–4
  Weber Blue, 12
Agave americana, 15
  angustifolia, 6, 15
  asperrima, 15
  atrovirens, 15
  cocui, 15
  complicate, 15
  crassispina, 15
  ferox, 15
  gracilipes, 15
  hookeri, 15
  inaequidens, 15
  lechuguilla, 15
  milliflua, 15

  potatorum, 15
  salmiana, 15
  tequilana, 2–16
  weberi, 15
agave moth, 16
agave snout weevil, 14, 16
Agriculture Department, U.S., 70
aguamiel, 2
aguardiente, 68, 102, 131
Agwa liqueur, 155
albedo, 299
alder, 234–5
ale, 211
alembic pot still, 24
Aleppo pine, 250
Alexander the Great, 96
alkermes, 153
allasch, 150
allspice, 136–38
almond, 260, 262, 307–8
aloe, 138–39, 157
Aloe vera, 138–39
aloin, 139
Aloysia citrodora, 322
  triphylla, 175–77
Alpha calendula, 328
alpha acids, 213
alpine strawberries, 347, 350
Alpinia galangal, 162
  officinarum, 162
Altissima rum, 106
Amanita muscaria, 233
amaretto, 153, 260, 262
Amaretto di Saronno, 260, 308
Amaro Averna, 163
Amaro del Carciofo, 143
amarogentin, 163–64
*amaro*s, 139, 143, 153
Amarula Cream, 121
Ambrosia melon, 346
Amer Picon liqueur, 275
American Agriculturist, 92
American Gin Company, 278
American Medical Association, 245
American white oak, 55
ammonium, 33

Amomum subulatum, 151
amygdalin, 260
amylase, 115
Anacardium occidentale, 113
Ananus comosus, 347
Anderson, James, 84–85
anethole, 178–84
Anethum graveolens, 322
Angel Red, 335
Angelica, 140–41, 200, 322
Angelica archangelica, 140
angels' share, 51
angostura, 228–33
    bitters, 163
Angostura trifoliata, 228–33
animal intoxication, 132, 246
anise, 179, 200
    star, 183, 200
anise hyssop, 180, 322
Anti-Saloon League, 261
Aperol liqueur, 163
Apium graveolens, 346
apple, 17–30, 334
    bugs in booze, 29
    Calvados, 17, 22–24
    cashew, 113–14
    cider, 18–22
    genetics of, 17
    grow your own, 18
    spirits made from, 22–23
    Vavilov Affair, 25
    Washington and, 23–24
    yeast and, 26–28
apple brandy, 22
apple liqueur, 22
apple wine, 22
applejack, 22, 23–24
apricot, 260–62, 307, 308, 334
aqua vitae, 34
aquavit, 149–50
Arabian jasmine, 218
Arapaho blackberry, 342
Araucania araucana, 122–23
araucania, 122–23
Arbequina olive, 335
Arbutus unedo, 131–32
Arctander, Steffen, 148
Ardbeg Scotch, 40
Aristotle, 241
Armagnac, 68
arolla stone pine, 250
aromatized wines, 63, 65

Arp rosemary, 323
arrack, 102, 119, 186
Art in the Age of Mechanical
    Reproduction distillery, 192
Art of Cookery Made Plain and Easy
    (Glasse), 255–56
Artemisia absinthium, 66, 200–3
    capestris, 203
    genipi, 203
    glacialis, 203
    pontica, 203
    rupestris, 203
    umbelliformus, 203
artichoke, 142–43
Artocarpus heterophyllus, 120
arzente, 68
ascorbic acid, 256
Asparagales, 2
Aspergillus oryzae, 79–80
asters, 142–43, 159, 200–203, 205
Aureus hops, 216, 342
Austen, Jane, 255
Australian Herbal Liqueur, 246
autumn crocus, 223
Averell Damson Gin Liqueur, 278
Averna Amaro, 314
Aviation, 225, 226
Aviation gin, x, 191
avocados, 118, 144
Azande tribe, 117
Aztec tribe, 352

β-asarone, 148
β-phellandrene, 140
bacanora, 10, 15
Bacardi rum, 94
bacteria, 3–4, 119
bagasse, 101
baijiu, 88, 90, 94
Bajtra liqueur, 126
bamboo wine, 130
banana, 112
Banks, Joseph, 255
Bärenjäger liqueur, 117
bargnolino, 284
barley, 31–41
    botany of beer, 32, 34
    breeding a better grain, 39
    domestication of, 32
    fertilizer and, 35, 37
    grow your own, 36
    growing the perfect, 34–37

malted, 31, 37, 39
   spring or winter, 35
   two-row or six-row, 34–35
Bartlett pears, 30
basil, 322
Batavia Arrack, 102, 186
Baudoinia compniacensis, 51
bay laurel, 144
bay rum, 137
Bayer, 220
beans, 133, 160–61, 182, 197, 252–54
Bearss lime, 298
Beaux Arts Fizz, 338
Bee Vodka, 117
beer
   ale and lager, 211
   banana, 112
   botany of, 32, 34
   brown bottles for, 212
   cassava, 115–17
   color of, 33
   hops and, 210–17
   invention of, 32
   psychoactive, 87
   rice and, 76, 82
   sorghum, 90–92
   wheat, 107–10
Beerenburg, 144
bees, 116–17
Ben Lomond currant, 266, 342
Bendistillery, 172
Benedictine, 86, 138, 140, 174, 188, 199, 224
benzyl acetate, 218
Beowulf Braggot, 117
Berlin Decree, 105
Bernard Loiseau Liqueur de Poires Laurier, 144
Bernhardt, Sarah, 154
berries, 340–44
betel leaf, 145–46
Betula papyrifera, 234–35
betulinic acid, 235
Bianca hops, 216, 342
Big Mackey plum, 335
bin Laden, Osama, 252
birch, 234–35, 312–13
birthworts, 157
bison grass, 146–47
Bison Grass Cocktail, 147
bitter almond, 307
bitter orange, 287–91, 300, 335

bittersharp apples, 21
bittersweet apples, 21
black currant, 263–67, 342
Black Diamond blackberry, 342
Black Gold, 75
Black Lace elderberry, 208, 328
Black Nail, 38
black pepper, 157
Black Republican cherry, 275
black spruce, 256
blackberry, 342
blackthorn shrub, 281–84, 342
Blake, Michael, 43
Blakeney Red pear, 30
blended bourbon, 48
blended whiskey, 48
blessed thistle, 143
blood orange, 301, 335
Blood Orange Sidecar, 301
Bloody Mary, 74, 133
blueberry, 342
Blue Fortune anise hyssop, 322
blue gum eucalyptus, 246
Blushing Mary, 354
Board of Food and Drug Inspection, 272
Bols distillery, 222
Bombay Sapphire gin, 168
Bonal Gentiane Quina wine, 61, 65
Bonnie Prince Charlie, 38
Bonpland, Aimé, 228
Booker's Bourbon, 40
borage, 328
botrytis bunch rot, 58
boukha, 271
bourbon, 33, 48
   birth of, 45–47
   garden cocktails with, 348
   oak barrels and, 53–54
   perfect corn for, 50
   wheat and, 109
bourgeon de sapin, 250
Boyle, Robert, xii
bracken ferns, 187
braggot, 117
brandy, 262
   invention of, 60–63
   types of, 68
brandy de jerez, 68
breadfruit, 120
brem, 81
British myrrh, 184

Brooklyn Cocktail (hybridized), 275
broomcorn, 89, 92
brown sugar, 100
brucine, 158
buckwheat, 108
buckthorn, 248
Buddha hand citron, 293
Budweiser, 76, 82
Buena Vista's Irish Coffee, 311
bugs in booze
    agave snout weevil, 16
    alkermes scale, 56
    cochineal, 129
    earthworms, 41
    honeybees, 116–17
    yeast, 29
bullace plum, 277, 278
Burbank, Luther, 277
burnet saxifrage, 179
"burnt wine," 60
Byrrh, 240

C. W. Abbott & Co., 230, 231
cabbage roses, 221
Cabernet Sauvignon, 64
cacao, 268–69, 313–14
cachaça, 101, 102
cactus, 126–28, 138
Café des Beaux Arts, 338
caffeine, 309–11
Caipirinha, 102
calamansi, 291
calamondin, 291, 294, 335
calamus, 148
calendula, 328
Calhoun, Scott, x-xi
California bay laurel, 144
Calisay, 240
Callen, Eric, 3, 5
Calrose rice, 78
Calvados, 17, 22–24, 138, 247
Calvados Domfrontais, 23
Calvados Pays d'Auge, 23
Campaign for Real Ale, 211
Campari liqueur, 129, 148, 163, 236, 292
cannabis, 210–17
Cantimpré, Thomas von, 169
caowy, 73
capillaire syrup, 185, 187
capsaicin, 353
Capsicum annuum, 346
    frutescens, 346

caramel, 33, 40
caraway, 149–50
carbon dioxide, 26, 29
carcinogens, 145, 148, 187, 192
Cardamaro Vino Amaro, 143
cardamom, 151–52, 157
cardoons, 142–43
Caribou, 258
carmine dye, 129
Caroline raspberry, 342
carrots, 124–25, 140, 149–50, 156, 179, 180, 181, 184
Carum carvi, 149–50
Cascade hops, 216
cascarilla, 236
cashew apple, 113–14
cashews, 113–14, 121, 190, 247
cassareep, 116
cassava, 115–17
Cassell's Dictionary of Cookery, 125
cassia cinnamon, 242
cassis, 263–64
Cassis de Dijon case, 267
cayenne pepper, 353
celery, 346
Centaurium erythraea, 160
Centers for Disease Control and Prevention, 245
century plants, 2
cereal crops, 31, 107
Cerro Baúl brewery, 190
chamazulene, 205
Chamaemelum nobile, 205
Chambord liqueur, 300
chamomile, 205
Champagne Cocktail, 229, 247
Chapeau Banana lambic, 112
Chapman, John, 20
charanda, 102
Charbay Distillery, 317
Chardonnay, 64
Charentais melon, 346
Charles Jacquin et Cie distillery, 300
Chartreuse, 138, 140, 148, 174, 223
    distillery, 278–79
Chateau Jiahu beer, 76
Cherry Bomb pepper, 346, 352
cherry brandy, 274
Cherry Heering, 274
cherry laurel, 144
Cherry Pick pepper, 346, 352
cherry tree, 276, 334
cherry wine, 274

Chibuku beer, 91
Chibuku Shake-Shake, 91
chicha, 42–45
chicha de jora, 48
Chichimeca people, 126
Chilean strawberry, 350–51
chili pepper, 352
chill filtration, 40
China Martini wine, 240
Chinese mandarin, 300
chinotto, 292
Chippewa blueberry, 342
chocolate, 268–69
Chopin vodka, 74
Chou En-lai, 95
Ciao Bella, 295
cider, apple, 18–22
    alcohol content of, 21
    ancient makers of, 19–20
    classification of, 21–22
    hard, 19
    pear and, 30
    preserving heritage, 19
    water and, 21
    yeast and, 26
Cider Cup, 20
cider gum eucalyptus, 246
cider sickness, 4
cilantro, 156, 322
cinchona, 237–40
Cinnamomun aromatic, 242
    verum, 241–42
cinnamon, 241–42
Ciroc vodka, 69
citral, 174
Citrofortunella microcarpa, 291
citron, 285, 286, 293, 335
citronellal, 174
Citrullus lantatus, 347
citrus, 151, 153, 285–305
    anatomy of, 299
    essential oils, 291
    grow your own, 294
    Meyer and, 296, 297
    orange liqueurs, 305
    peels, 293
    sprayed, 290
    zester, 293
    see also specific citrus
Citrus aurantifolia, 298, 334
    aurantium, 287–92, 335
    bigaradia, 288
    hystrix, 298

× *junos,* 302
latifolia, 298
limon, 297, 334
maxima, 300
medica, 293
nobilis, 300
× *paradisi,* 295
reticulata, 300
sinensis, 301
unshiu, 300
Clark, John B., 41
Classic Margarita, 8
Classic Martini, 173
Claviceps purpurea, 86–87
Claytons Kola Tonic, 314
Clear Creek Distillery, 244
clementine, 300
cloves, 136, 152–53, 242
coca, 154–55
Coca-Cola, 155, 199, 314
cocaine, 154–55
Cocchi Americano wine, 65
cochineal scale, 56, 129
cocktail glasses, xvi
cocktails
    recipes, viii
    see also specific cocktails
coconut tree, 119
Coffea arabica, 309–11
    canifora (robusta), 309
coffee, 248, 309–11
Cognac, 60, 219
Cointreau liqueur, 305
Cola acuminata, 313–14
Colchicum autumnale, 223
Colglazier & Hobson Distilling, 94
colonche, 128
Colubrina arborescens, 248
    elliptica, 248
Columbus, Christopher, 42, 43, 97,
    286, 352
Comadia redtenbacheri, 16
Combier, Jean-Baptiste, 287
Combier Distillery, 195, 249, 287
Combier liqueur, 305
Commiphora myrrha, 249
Complete Distiller, The (Cooper), 218
confectio alchermes, 56
congeners, 27–28
Cook, Captain James, 255
Cooper, Ambrose, 218
Cooper Spirits International, 225
coopers, 52–54

coppicing, 241, 312
coriander, 156, 171, 200
Coriandrum sativum, 322
Corinto cucumber, 346
cork oak, 51
corn, 42–51, 89
    birth of bourbon, 45–47
    chewing uncooked, 44–45
    chicha, 42–45
    choosing the perfect, 50
    glass of, 48–49
    sex life of, 44
    types of, 46
corn beer, 42–45, 48–49
corn vodka, 49
corn whiskey, 45, 47, 49
Cornell University, 19, 24, 265, 335
cornstalk wine, 42–45
Corona Extra, 49
Corylus avellana, 312–13
    maxima, 312
coumarin, 146–47, 194, 197, 242
"country wines," 124
Creasy, Lara, 251
cream, 265
crème, 265
crème de, 265
crème de cassis, 263–67
crème de figue, 271
Crème de Noisette, 313
crème de noyaux, 262
crème de violette, 225
crème Yvette, 225
Crespigny, Rose Champion De, 284
Crispin's Rose Liqueur, 222
Crocus sativus, 223
Croton eluteria, 236
cubeb, 157
cucumber, 346
Cucumis melo, 346
    sativus, 346
Culpeper, Nicholas, 185, 220
cumin, 150
Curaçao liqueur, 287–88, 305
currants, 263–67, 342
curry, 146
Cusparia febrifuga, 228
    trifoliata, 228
cyanide, 115, 208, 260, 307, 327
Cymbopogon citrates, 322
Cynar, 143
Cynara cardunculus, 142
    scolymus, 142–43
cynarin, 142
cynaropicrin, 142
cypress, 169–73

daiginjo, 82
Daiquiri, 99
Daktulosphaira vitifoliae, 64, 66
damask roses, 221, 222
damiana, 158
damson plum, 277–78, 281, 335
dandelion, 143
Dark and Stormy, 167
Dark Northern rye, 85
darnel, 84
Dasylirion wheeleri, 10
date palm, 118–19
de Garine, Igor, 81–82
De Materia Medica (Dioscorides), 56, 96
Death's Door distillery, 172
Delessert, Benjamin, 105
demerara sugar, 100
Denominación de Origen, 10
dent corn, 46
dépense, 20
Der Naturen Bloeme (Maerlant), 169
Destillerie Purkhart, 225
dill, 322
dioecious plants, 57, 122, 169, 171, 213
Dioscorides, 56, 96
Distell distillery, 121
distilled gin, 170
Distilled Spirits Council, 85
Distillerie des Menhirs, 108
Distillerie Miclo, 222
distillers, 26–28, 31
    condensing techniques, 41
dittany of Crete, 159
DO, 10, 11
Dogfish Head Brewery, 44–45, 76, 117, 223, 269
Dolin Blanc Vermouth de Chambéry, 65
Domaine de Canton liqueur, 166
Dombey, Joseph, 175, 177
Dombey's Last Word, 175
Douglas, David, 244
Douglas Expedition, 244
Douglas fir, 243–44
Dr. Struwe's Suze and Soda, 164
dragon fruit, 128
Drambuie liqueur, 38, 140, 146
Drosera rotundifolia, 193–94

drupe, 308
Dukat dill, 322
Dutch courage, 172
Dwarf Cherry nasturtium, 329
dyes, 129, 290

earthworms, 41
eau-de-vie, 22–23, 24, 61, 63, 69, 243,
    244
eau-de-vie de prunelle sauvage, 284
eau-de-vie de sorbier, 281
Ebers Papyrus, 200
Eddu Silver whiskey, 108
Edmond Briottet distillery, 313
egg whites, xvii
eglantine rose, 222
Elder, Joel, 50
elderberries, 208
elderflower, 206–9, 328
    imbibing, 209
Elderflower Cordial, 207
Elderflower-Lavender Champagne
    Cocktail, 330
elecampane, 159
Elettaria cardamomum, 151–52
Elfin King strawberry tree, 132
Ellestad, Erik, 251
Emerson, Edward Randolph, 73
Emma (Austen), 255
endocarp, 299
English Physician, The (Culpeper), 221
Enterolobium cyclocarpum, 7
enzyme, 27, 79–81, 115, 123
ergot, 86–87
Erythroxylum coca, 154–55
    novogranatense, 155
    rufum, 155
estragole, 180
Estratto di Eucaliptus liqueur, 245
ethyl alcohol, 26–28
ethylene, 290
Eucalittino delle Tre Fontane liqueur,
    245
eucalyptol, 144, 151, 180
eucalyptus, 245–46
Eucalyptus citriodora, 245
    globulus, 246
    gunnii, 246
Eucalyptus Gum Leaf Vodka, 246
eugenol, 136, 137, 153, 242
Eureka lemon, 297, 334
European centaury, 160
European mountain ash, 281

European oak, 55
European Union (EU), 129, 148, 247, 267
Evanow, Allison, 86
Evelyn, John, 234
exocarp, 299
"Eyaws," 41

Faerie watermelon, 347
Fagopyrum esculentum, 108
falernum, 298
Falimirz, Stefan, 72
false cocaine, 155
Fantin-Latour roses, 221
farnesol, 218
Federal Alcohol Administration, 54
Fee Brothers, 229
Fejérváry-Mayer, 2
Femminello Ovale lemon, 297
feni, 113, 114
fennel, 180–81, 200, 322
fenugreek, 146, 160–61
fermentation, 26–29, 59
    barley and, 31
    date palm, 199
    rice and, 79–81
fern, 185–87
Fernet Branca liqueur, 139, 224, 246,
    249, 250
Ficus carica, 270–71, 334
fig, 270–71, 334
filbert, 312
filé, 192
Filipendula ulmaria, 187
Filipino still, 7–8
Finissimo Verde basil, 322
fire blight, 30
Firefly distillery, 310
flavedo, 299
flint corn, 46
flor, 61
Florence fennel, 180, 322
flour corn, 46
flowers, 204, 327–31
    see also specific flowers
fly agaric, 233
Foeniculum vulgare, 180, 322
Food and Drug Administration (FDA),
    139, 145, 148, 155, 162, 187, 192, 197,
    246, 256, 273
Food & Function, 146
Forbidden Fruit, 300
Fosberg, Raymond, 239
Founding Fathers, 45, 63–64

Four Winds Growers, 296
fractional column still, 24
Fragaria × *ananassa*, 347, 350–51
  chiloensis, 350
  moschata, 350
  vesca, 350
  virginiana, 351
François, Antoine, 300
Frangelico liqueur, 312
Frank Meyer Expedition, 297
Franklin, Benjamin, 43, 255–56
Fratello liqueur, 312
Fred Boutin lavender, 328, 330
freeze distillation, 24
French Intervention, 13
French oak, 55
Freud, Sigmund, 154
Frézier, Amédée François, 350–51
Frézier Affair, 351
From Honey to Ashes (Lévi-Strauss), 195
Frontenac grape, 67
fruit, 259, 345–54
  see also specific fruit
"fruit bricks," 59
Fryar, Alan, 47
Fuggle hops, 216
fungi, 58, 79, 84, 86–87, 233
fusarium, 39

Gabriel Boudier distillery, 144
Gaia Estate winery, 250
galangal, 162
Galen, 169
Galipea officinalis, 228
  trifoliata, 228
Galium odoratum, 194
Galliano liqueur, 140, 179, 199
Gamache, Darren, 215, 217
gamju, 81
garden-infused simple syrup, 324
Garey's Eureka lemon, 297
génépi, 203
genever, 170
Genovese basil, 322
genshu, 82
gentian, 160, 163–65, 309, 313
Gentiana lutea, 163–165
Gentiane liqueur, 163
gentiopicroside, 163–64
Gentry, Howard Scott, 6
George VI, King, 220
geraniol, 174
geraniums, 323

geranyl acetate, 156
Gerard, John, 307
German chamomile, 205
germander, 165
Giffard Pamplemousse liqueur, 295
gin, x–xi
  common ingredients of, 171
  garden cocktails with, 348
  grape-based, 69
  juniper and, 169–73
  types of, 170
gin and tonic, 240
  Mamani, xi, 238
ginger, 151–52, 162, 166–67, 168
ginger beer, 166–67
ginjo, 82
ginseng, 155
Glasse, Hannah, 255–56
glasses, xvi
glycyrrhizin, 182
Godfrey's Cordial, 192
Godin Tepe site, 32, 34
Gohyakumangoku, 78
Gold Flame honeysuckle, 328
Golden Jubilee anise hyssop, 322
Goldschläger liqueur, 242
Gómara, Francisco López de, 4
gomme syrup, 252, 253
gooseberry, 263–67
goosefoot, 105–6
Gordal olive, 335
gorillas, 168
Goschie, Gayle, 215
Gosling's, 167
Gouais Blanc, 64
Gouberville, Gilles de, 22
grains, 31
  see also specific grains
grains of paradise, 168
Grand Marnier liqueur, 288, 289, 305
grapefruit, 295, 300
grapes, 57–69, 310
  fungus on, 58
  invention of brandy, 60–63
  phylloxera and, 64, 66
  spirits made from, 68–69
  wine, *see* wine
  yeast and, 59
Graphic, 303
grappa, 205
grasses, 31–56, 76–104, 107–10, 130,
  146–47
greater galangal, 162

green mint, 325
green tea, 218
GreenBar Collective distillery, 219
greenbriar, 191
greengage plum, 277–79, 335
Greenwood, James, 170
Grenada pomegranate, 335
grenadine, 337–39
    homemade, 338
Grey Goose vodka, 109
Grosso lavender, 328, 330
grow your own
    apples, 18
    barley, 36
    black currants, 266
    cherry tree, 276
    citrus, 294
    elderberries, 208
    hops, 216
    lemon verbena, 176
    sloes, 282
    wormwood, 202
guarana, 155
guignolet, 274
gum arabic, 252–54
Gum Arabic Company, 252
gusano, 16
G'Vine gin, 69

Haig, Alexander, 88
Hangar One Vodka, 69, 298, 300
Hansen, Johnnie, 261
Harbin Beer, 49
Haus Alpenz, 225, 317
hazelnut, 312–13
heath, 131–32
Hefeweizen, 109
heirloom corn, 50
Helmont, Johannes Baptista van, 234
Hemidesmus indicus, 191
Hendrick's Gin, 205, 222
Henri Bardouin, 197
herb hyssop, 181
*Herball, or Generall Historie of
    Plantes, The* (Gerard), 307
herbs, 135, 320–26
    growing notes, 322–23
    spanking your, 324
    see also specific herbs
hermaphrodite plants, 57, 67
heroin, 219, 220
hesperidium, 285
heterozygosity, 17

Hibiki whiskey, 48
Hidcote lavender, 328, 330
Hidden Marsh distillery, 117
Hierochloe odorata, 146–47
highball glasses, xvi
Highland Park Scotch, 33
Hildegard, St., 263
Hilltop Baldwin currant, 266
Historia Animalium, 241
Historia Plantarum (Ray), 136
Historias y Sabores distillery, 196
Holt's Mammoth sage, 323
Homer, 117, 219–20
Honey Drip, 93
honeybees, 116–17
honeyjack, 117
honeysuckle, 206–9, 328
hop kiln, 215
hop marjoram, 159
hops, 210–17, 342
    grow your own, 216
    international bitterness units, 214
    kiln, 215
    varieties of, 214
Hordeum vulgare, 31–41
Humulus japonicus, 210–17
    lupulus, 210–17, 342
Hutchinson, Horace, 284
hydrangea, 327
Hypopta agavis, 16
hyssop, 181
Hyssopus officinalis, 181
Hysterical Water, 255

ice, xvii
Ichang Papeda, 295, 302
Illicium anisatum, 183
    verum, 183
Imbibe magazine, 251
Improved Meyer lemon, 294
Incan treasure, 71–72
Indian corn, 42
Indian sarsaparilla, 191
injera, 89
"Instructions to a Painter," 172
International Bartenders Union, 261
International Exhibition of 1862, 310
International Plant Names Index
    (IPNI), 12
Inula helenium, 159
ionone, 226
Ipomoea batatas, 73
iridoid glycosides, 160

Iris germanica, 189
  pallida, 189
irises, 189, 223–24
Irish coffee, 311
Irish Mist, 38
irone, 189
ironwood, 248
Iroquois tribe, 257, 258
Islay Scotch, 40
isohumulones, 212
Iznik cucumber, 346

J. Witty Spirits, 205
Jack Daniels whiskey, 54
Jack Rose, 339
jackfruit, 120
Jacques Cardin distillery, 219
jaggery, 119
Jam Session plum, 335
Jamestown colony, 42
Japanese cherry trees, 276
Japanese oak, 55
jasmine, 218–19, 328
Jasminum officinale, 218–19, 328
Jefferson, Thomas, 45, 64, 257
Jerry Thomas' Regent's Punch, 185, 186
jigger, xvi
Jinro, 81, 94
Johnny Appleseed, 20
Johnny-jump-up, 329
Johnson, Edward, 124
Jonge, 170
Jonkheer Van Tets currant, 266
Joy perfume, 219
Jubilejná Borovička, 172
Juglans regia, 315–17
juniper, 169–73, 200
Juniperus communis, 169–73
junmai, 82

Kaempferia galangal, 162
Kaffir lime, 298, 334
Kahlúa liqueur, 199, 310
Kalmia latifolia, 144
Karlsson, Börje, 75
Karlsson's Gold vodka, 75
kasha, 108
Kaufman, Henry F., 158
Kentucky bourbon, 45, 47
Kentucky Bourbon Trail, 47
Kentucky Colonel mint, 323, 325
Kentucky Common Beer, 49
Kermes, 129

Ketel One vodka, 109
key lime, 298, 334
Kilchoman distillery, 37
King's Ginger liqueur, 166
Kir, 264, 267
Kir, Félix, 264
Kirin beer, 82
kirsch (kirschwasser), 274
Kissinger, Henry, 95
knotwood, 108
koji mold, 79–80
kola nut, 313–14
koshu, 82
Koval distillery, 219, 222
kümmel, 150
kumquats, 335
kykeon, 117

lactobacillus, 47
Lagenaria vulgaris, 3
lager, 211
Laird, Alexander, 23
Laird & Company Distillery, 23
Lakang Hari Imperial Basi, 102
lambic brewers, 29
lao-lao, 82
Laphroaig distillery, 37, 54
Laraha orange, 287–88
Laricifomes officinalis, 233
Last Word, 175
laudanum, 220
laurel, 144, 192, 241–42
Laurus nobilis, 144
Lavandula angustifolia, 330–31
  × *intermedia,* 330
lavender, 151, 328, 330–31
Lavender-Elderflower Champagne
  Cocktail, 330
Lavender Martini, 331
Lavigerie, Charles, 154–55
Lazaro de Arregui, Domingo, 8
Ledger, Charles, 239
lemon, 294, 297, 334
lemon balm, 174
lemon verbena, 175–77, 322
lemongrass, 322
Leuconostoc mesenteroides, 4
Lévi-Strauss, Claude, 195
Lewelling, Seth, 273, 275
Liber de Natura Rerum (Cantimpré),
  169
licor de cocuy, 15
licor de madroño, 131

Licor de Pomelo, 295
licores de tamarindo, 133
licorice, 178–84
  list of drinks flavored by, 182
Lightning Pear, 30
Lillet wine, 61, 65, 240, 331
lime, 286, 294, 297, 298, 334
limestone water, 47
limoncello, 293, 344
limonene, 140, 157, 171, 180, 299
linalool, 144, 151, 156, 174, 218, 242, 298
linalyl acetate, 151
Lincoln County process, 54
Lindley, John, 246
Linie Aquavit, 149
Linnaeus, Carl, 200
liqueur, 265, 267
liqueur de noix, 317
Liqueur de Pain d'Epices, 242
Liqueur Herbert, 162
Liquore Elixir di China, 240
Liquore Strega, 141
Lisbon lemon, 334
Litchi chinensis, 334
Little Baby Flower watermelon, 347
Lolium temulentum, 84
London gin, 170
Lonicera × heckrottii, 328
loosestrife, 337–39
Lophophora williamsii, 7
Los Angeles Times, 145–46
louche, 178, 201
Louis-Frères, Simon, 19
LSD, 87
lupulin, 213
Luxardo Amaretto di Saschira
    Liqueur, 308
Luxardo Company, 272, 273, 274
lychee, 334

mabioside, 248
mace, 188
madder, 194, 237–40, 309–11
madeira, 62, 63
madrone, 131–32
Maerlant, Jacob van, 169
mahon, 170
maidenhair fern, 185–87
maize, 42
maguey, 2, 3
Mai Tai, 304
Maker's Mark bourbon, 54, 109
Malabar cardamom, 151

malaria, 237–40, 245
mallow, 268–69
malting, 31, 37, 39
  green malt, 37
Malus domestica, 17–30, 334
Mamani, Manuel Inra, 238, 239
Mamani Gin & Tonic, xi, 238
mandarin, 285, 297, 300, 301, 302
Mandarine Napoleon liqueur, 300, 305
mango, 190
Manhattan, 86, 163
Manihot esculenta, 115–17
manioc root, 115
mao-tai, 88, 94–95
maple, 257–58
maple sugar, 257–58
marasca cherry, 271–76
maraschino, 274
maraschino cherry, 271–76, 334
  homemade, 273
March violet, 225
Margarita, 126, 127, 133
  Classic, 8
marigold, 328
marijuana, 213
Marker, Russell, 191
Marolo Distillery, 205
Marquette wine, 67
Marsala, 62, 63
Martini, 173
marula, 121
Master of Malt, 314
mastic, 247
Mastika, 247
Matricaria chamomilla, 205
  recutita, 205
mauby, 248
Mauricia Tamarind Liqueur, 133
Maurin Quina liqueur, 240
May wine, 194
Mayflower, 210, 213
McCarthy, Stephen, 243–44
McGovern, Patrick, 32, 44, 60, 76, 223
McKay, Steven, 265
mead, 116–17
meadowsweet, 187
Meinhard, Dr. Teodoro, 230
Melissa officinalis, 174
melon, 346
Mentha spicata, 323, 325
Menzies, Archibald, 122
Meredith, Carole P., 64
Merwin, Ian, 24

mescal, 5
    see also mezcal
mescaline, 7
mesocarp, 299
metaxa, 68
methyl salicylate, 235
Mexican lime, 298, 334
Meyer, Frank N., 296, 297
Meyer lemon, 294, 296, 334
mezcal, 4–16
    bugs in booze, 16
    field guide to, 11
    list of spirits, 15
    tasting, 14
Micheladas, 133
Michelberger 35% liqueur, 197
Midas Touch beer, 223
mijiu, 81
mildew, 39
millet, 89
Minnesota 13 corn, 50
mint, 159, 165, 174, 180, 324, 330–31
Mint Julep, 325, 326
mirabelle plum, 277, 279, 335
miracle fruit, 346
mistelle, 65
Mister Lincoln rose, 329
mixtos, 10, 11
Miyama Nishiki rice, 78
mobbie, 73
Mojito mint, 323, 325
Mojito y Mas, 104
mojo, 287
molasses, 101, 102, 106
Molasses Act of 1733, 101
Mongozo Brewery, 112
Monin Original Lime liqueur, 298
monkey puzzle, 122–23
monkey rum, 93
Moor Park apricot, 260
Moore, Marianne, 122
moonshine, 28
    corn, 47, 49, 50
    sorghum, 92–93
morning glory, 73
Morris, Chris, 50
Moscow Mule, 74, 167
mountain laurel, 144
Mourt's Relation, 210, 213
Moutai, 94
Moxie, 165
Mud Puddle vodka, 269
mudai, 123

muddling, xvii
mulberries, 120, 270–71
Munstead lavender, 328, 330
Musa acuminata, 112
muscatel (moscatel), 62
muscovado sugar, 100
musk lorikeet, 246
musk strawberry, 350
musto verde, 69
myrcene, 171
Myristica fragrans, 188
myrrh, 249
myrtles, 136–38, 152–53, 245–46
Mysore cardamom, 151

Nabhan, Gary Paul, 6
nama, 82
Napoleon Bonaparte, 105
Napoleon III, 287
nasturtium, 329
Natural History (Pliny the Elder), 83
navai't, 128
naval spirit, 103–4
Navelina orange, 301
Navel orange, 301, 335
nectarines, 335
Negroni, 143, 163, 164, 292, 295
Nemiroff Birch Special Vodka, 235
Neon calendula, 328
neroli oil, 291, 302
neutral spirits, 106
New Deal Distillery, 269
*New Forest: Its Traditions, Inhabitants
    and Customs, The* (Crespigny), 284
New York Times, 70, 76, 338
Nicotiana tabacum, 195–96
nicotine, 195, 196
nightshades, 70–72, 74–75, 195–96,
    352–53
nigori, 82
Nipah virus, 119
Nixon, Richard, 88, 94, 95
noble rot, 58
nocino, 315–17
    homemade, 316
Nocino della Cristina liqueur, 317
Noir de Bourgogne currant, 266
nopales, 126
Northern United Brewing Company,
    106
Nostalgie liqueur, 317
Noyau de Poissy distillery, 262
nutmeg, 157, 188

nuts, 306
    see also specific nuts
Nux Alpina Walnut Liqueur, 317

oak, 52–54
    cork, 51
    field guide to, 55
oast house, 215
Ocimum basilicum, 322
Odyssey, The (Homer), 117, 219–20
oil of lemon eucalyptus, 245
Old-Fashioned, 25, 48, 163
Old-Fashioned glasses, 16
Old Sugar Distillery, 94, 106
Old Tom gin, 170
Olea europaea, 335
olive, 218–19, 335
    brine for, 336
Omachi rice, 78
On Herbs and Their Powers (Falimirz),
    72
100% blue agave spirits, 15
opium poppy, 219–20
Opuntia, 126–29
    humifusa, 126
orange, 286, 300, 335
    liqueurs made from, 305
orange flower water, 302
Orangerie liqueur, 301, 305
orchid, 198–99
O'Rear, James, 47
oregano, 159
Oregon myrtle, 144
Oregon oak, 55
orgeat, 304
Origanum dictamnus, 159
orris, 189
Oryza sativa, 76–82
oude, 170
ouicöu, 115
Oviedo, Fernandez de, 129
Oxytenanthera abyssinica, 130

paan, 145–46
pacharán, 284
paciki, 49
Pagès Védrenne distillery, 177
Pálinka, 222
palm, 118–19
Panama-Pacific International
    Exposition, 94
pansies, 329
Papaver somniferum, 219–20

paper birch, 234
paprika, 353
Parfait Amour liqueur, 222, 225
Parker, Lucinda, 244
Parma violets, 226
parsnip, 124–25
"Passing of the Eucalyptus," 245
Pastinaca sativa, 124–25
pastis, 106, 178–184
    Perfect, 181
patxaran, 182, 284
Paul Devoille Distillery, 242
peach, 307, 335
pear, 30
pear brandy, 30
pear cider, 30
peat-smoked barley, 37, 41
pechuga mescal, 9
Peguis jalapeño, 346
Pehuenche people, 123
Pelargonium, 323
pellagra, 89
Pemberton Distillery, 74–75
pentosan, 83
Peppadew pepper, 352
pepper, 346, 352–53
peppers, 145–46, 157
Percy, Walker, 326
Perfect Distiller and Brewer, The, 72
Perfect Pastis, 181
Perfection fennel, 322
Perique Liqueur de Tabac, 195–96
perry, 30
Persian lime, 298
Pesto Perpetuo basil, 322
petitgrain oil, 291
peyote, 7
phenyl-acetic acid, 218–19
Philip, Prince, 104
Philosophical Works (Boyle), xii
phylloxera, 64, 66
Physalis philadelphica, 347
Picea, 255–56
    glauca, 256
    mariana, 256
    rubens, 256
pickles, 349
picrocrocin, 223
pie cherry, 334
pikake jasmine, 218
Pimenta dioica, 136
    racemosa, 137
pimento dram, 136, 138

Pimento-L pepper, 346
Pimm's Cup, 161, 328
Pimm's No. 1, 161
Pimpinella anisum, 179
piña, 5, 9
pine, 243–44, 250–51, 255–56
pineapple, 347
pineapple sage, 323
Pineau des Charentes wine, 65
pinene, 140
pink peppercorn, 190
Pinot Noir, 64
Pinus, 250–51
    cembra, 250
    halepensis, 250
Piper betle, 145–46
    cubeba, 157
    nigrum, 157
piperine, 157
piquanté pepper, 352
Pisco, 69
Pisco Sour, 67
Pistacia lentiscus, 247
pitahaya (pitaya), 128
pith, 299
Pizarro, Francisco, 71
plantains, 112
Plantlife UK, 173
Plat, Sir Hugh, 193
Pliny the Elder, 83, 85, 163, 325
plum, 277–79, 335
Plymouth distillery, 28
Plymouth Gin, 170, 283
pod corn, 46
poet's jasmine, 218–19
Poire Williams brandy, 30
poison ivy, oak, and sumac, 113, 121, 190
Polka Red raspberry, 342
Polly, 338
Polmos Białystok distillery, 147
POM Wonderful, 335
pomace brandy, 69
pomegranate, 319, 335, 337–39
pomelo, 285, 286, 295, 297, 300, 301
pommeau, 23
pony, xvi
popcorn, 46
poppy, 219–20
port, 62, 63
Portuguese oak, 51, 55
potato, 70–75
    artisanal, 74–75

birth of vodka, 72
Incan treasure, 71–72
sweet, 73
prickly pear cactus, 126–29
cochineal and, 129
sangria, 128
spirits from, 128
syrup from, 127
progesterone, 191
Prohibition, 50, 59, 213, 235, 261, 262, 272
prugnolino, 284
Prunus armeniaca, 260–62, 334
    avium, 272
    cerasus, 271–76, 334
    domestica, 277–79, 335
    dulcis, 307–8
    laurocerasus, 144
    mume, 279
    persica, 335
    salicina, 277
    serrulata, 276
    spinosa, 281–84, 342
    × *yedoensis,* 276
Pseudotsuga menziesii, 243–44
pulque, 2–4, 15, 45
punch au rhum, 102
Punica granatum, 335, 337–39
Punt e Mes wine, 61, 65
Pure Food and Drug Act, 158, 262
Pyrus communis, 30

quandong, 280
quebrantahuesos, 49
Queen Charlotte violet, 225
Queen Jennie Sorghum Whiskey, 94
Quercus, 51–55
    alba, 55
    garryana, 55
    mogolica, 55
    petraea, 55
    pyrenaica, 55
    robur, 53, 55
quetsche plum, 279
quid, 145
quinine, 237–40, 309, 313
Quinologie, 237
quinquina, 65

Raicilla, 10, 15
Rainier cherry, 275
raisins, 264
raki, 182, 271

raksi, 81
Ramos, Henry, 303
Ramos Gin Fizz, 302, 303
Raspail, François, 287
raspberry, 342
ratafia de cassis, 263
Rather, Dan, 88
raw sugar, 100
Ray, John, 136
"real whiskey," 33
recipes
  cocktails, viii
  egg whites, xvii
  glasses, xvi
  ice, xvii
  muddling, xvii
  serving size, xvi
  simple syrup, xvii
  syrups, infusions, and garnishes,
    ix
  tonic water, xvii
rectified spirits, 106
red currants, 266, 342
Red Lion Hybrid, 289
Redventure celery, 346
Refrigerator Pickles, 349
Reine Claude plum, 278
resurrection lily, 162
retsina, 250
Rheum rhabarbarum, 347
rhubarb, 347
rhum agricole, 101, 102
Ribena, 264
Ribes americanum, 266
  nigrum, 263–67, 342
  odoratum, 266
rice, 76–82
  as no ordinary grass, 76–77
  polishing, 78–79
  sake, 76–82
  spirits made from, 81–82
  varieties of, 78
Ritinitis Nobilis, 250
Rob Roy, 86
Rochefort, Henri, 155
Roman apricot, 260
Roman Beauty rosemary, 323
Roman chamomile, 205
Root liqueur, 192, 235
Rosa, 329
  centifolia, 221–22
  damascena, 221–22
  rubiginosa, 222

roses, 17–30, 187, 221–22, 260–61, 271–76,
    277–79, 281–84, 307–8, 350–51
rose hips, 222
rosemary, 323
Rose's Kola Tonic, 314
Rosmarinus officinalis, 323
rosolio, 193
rowan berry, 281
Royal, 48
Royal Ann, 275
Royal Combier, 249, 287, 305
Royal Horticultural Society, 244, 278
Royal Tannenbaum, 251
Royale pineapple, 347
Rubus, 342
  idaeus, 342
rue, 228–33, 285–305
rum, 73, 94, 102
  garden cocktails with, 347
  making, 99–103
  slavery and, 97
rum-verschnitt, 102
Rumona, 197
Rush, Dr. Benjamin, 257
Rusty Nail, 38
rye, 83–87
  comeback of, 85–87
  founding distiller, 84–85
ryegrass, 84

SABMiller, 91–92, 115–16
Saccharomyces cerevisiae, 4, 59, 61, 79
Saccharomycetales, 26–28
Saccharum barberi, 97
  officinarum, 96–106
  sinense, 97
saffron, 151, 198, 223–24
safranal, 223
safrole, 192
sage, 323
St. George Spirits distillery, 69, 298,
    300
St-Germain liqueur, 206, 207, 209
St. Nicolaus distillery, 172
St. Regis hotel, 338
sake, 76–82
  cocktail, 80
  enjoying, 79
  nomenclature of, 82
Sake One brewery, 78
salicylic acid, 158
Salvia elegans, 323
  officinalis, 323

sambuca, 182, 206
Sambucus canadensis, 208
   nigra, 206–9, 328
   racemosa, 208
Samuel Adams Summer Ale, 168
sandalwood, 280
sangria, 128
Santa Maria Novella, 189
Santalum acuminatum, 280
Santo cilantro, 322
sarsaparilla, 191, 235
sassafras, 192, 235
Satsuma mandarin, 300
Satureja hortensis, 323
   montana, 323
sauterne, 58
Savanna bamboo, 130
savory, 323
Savoy Cocktail Book (Craddock), 261
Sazerac, 184
scale, 56, 129
scented geranium, 323
Schinus molle, 49, 190
   terebinthifolius, 190
Schlehenfeuer, 284
Schramm, Tyler, 74–75
Scientific American, 8
Sclerocarya birrea, 121
Scottish Crops Research Institute,
   34, 39
Scotch whisky, 37–40, 86
   chill filtration and, 40
   water and, 40
scurvy, 103, 255, 256, 287
Scyphophorus acupunctatus, 14, 16
Secale cereale, 83–87
Seed, Eric, 225
seeds, 31, 118, 306
   see also specific seeds
Senegal gum tree, 252–54
Senegalia senegal, 252–54
Senior Curaçao of Curaçao liqueur,
   305
serving size, xvi
Sessile oak, 55
Seville bitter orange, 287–91
shaddock, 300
shandy, 166–67
sharp apples, 21
sherry, 61–63
shochu, 73, 76, 81–82, 108, 279
Sichuan button, 329
Sicilian lemon, 297

Siegert, Johannes G. B., 229–33
Siembra Azul distillery, 13
Sikkim Paan Liqueur, 145–46
Sikotok cloves, 153
Simmonds, Arthur, 278
simple syrup, xvii, 100
   garden-infused, 324
Sinistrari, Ludovico Maria, 157
Siputih cloves, 153
Sirionó tribe, 117
sirop de gomme, 254
slavery, 97, 257, 273, 275
slivovitz, 279
sloe berry, 281–84, 342
sloe gin, 170, 281–84
Sloe Gin Fizz, 283
Slow Bolt cilantro, 322
Smilax regelii, 191
Smirnoff vodka, 94, 167
Smith, John, 42
snag gin, 281
Snap liqueur, 166
soda and bitters, 230
soft drinks, 235, 248, 252
soju, 73, 81, 94
Solanum berthaultii, 71
   lycopersicum, 6, 347
   maglia, 71
   tuberosum, 70–75
soleras, 63
Solerno Blood Orange Liqueur, 301,
   305
Sonnentau Likör, 194
sonti, 81
sorbitol, 30
Sorbus aucuparia, 281
sorghum, 88–95
   American, 92–94
   beer made from, 90–92
   mao-tai, 88, 94–95
   as most-imbibed plant, 90
   survival of the fittest, 88–89
   sweet and grain, 89
   syrup, 92–94
   witchweed and, 95
Sorghum bicolor, 88–95
Sorgrhum, 94
Sorrento lemon, 297
sotol, 10, 15
sour cherry, 276
sour orange, 286, 301
Spacemaster 80 cucumber, 346
Spanish jasmine, 218

spearmint, 325–26
Species Plantarum (Linnaeus), 200
Spice Islands, 152
spices, 135
   see also specific spices
spilanthol, 329
spitters, 17
spitting, 44–45, 115, 123
Springbank distillery, 37
spruce, 255–56
spruce beer, 255–56
spurges, 115–16, 236
Square One vodka, 85
Stalin, Joseph, 25
star anise, 183, 200
star-vine, 183
Steenkamp, Johan, 352
Stepan Company, 155
straight bourbon, 48
straight rye whiskey, 85
straight whiskey, 33, 109
strawberry, 347, 350–51
strawberry tree, 131–32
Strega, 140, 141, 224
Striga, 95
Stroh 80 rum, 106
strychnine, 158, 232, 233
succade, 293
sugar, 100
sugar beet, 105–6
sugar maple, 257–58
sugarcane, 43, 73, 96–106
   birth of, 96–97
   botany of a cane, 98–99
   cultivars, 98
   guide to spirits, 102
   making rum, 99–103
   naval spirit, 103–4
   sugar primer, 100
sundew, 193–94
Sungold tomato, 347
Sunshine Flashback calendula, 328
Suntory distillery, 48
"supercassis," 263–64
superfine sugar, 100
Suro-Piñera, David, 13–14
Sutherland Gold elderflower, 328
Suze wine, 163, 164
swamp whiskey, 92
Swanston, Stuart, 34, 37, 39
sweet almond, 307
sweet apple, 21
sweet briar rose, 222

sweet cicely, 184
sweet corn, 46
sweet fennel, 180, 322
sweet flag, 148
sweet kernel apricot, 260
sweet orange, 286, 295, 301
sweet orange oil, 291
"sweet *paan,*" 145
sweet potato, 73
Sweet Success cucumber, 346
sweet violet, 226, 329
sweet woodruff, 194
sweetgrass, 146
SweetHeart apricot, 334
Sylva (Evelyn), 234
Sylvius, Franciscus de la Boë, 169
Synsepalum dulcificum, 346
Syzygium aromaticum, 152–53

Tagetes erecta, 328
Tahiti lime, 298
tahona, 9
Tamarind, 133
Tamarindus indica, 133
Tamborine Mountain Distillery, 246, 280
Tamiflu, 183
tangerine, 300
tannins, 53
Tantalus, 300
tapai, 81–82
Tapaus distillery, 295
tapuy, 81
Tarahumara tribe, 43
taxonomy, 6
tea, 146, 310
teff, 89
tejate, 49
tej (t'edj), 117
tejuino, 49
teosinte, 43
tequila, 8–10, 15, 133
   field guide to, 11
   garden cocktails with, 348
   protecting the plants, 13–14
   tasting, 14
Tequila Sunrise, 339
terpineol, 144
tesgüino, 49
Teucrium chamaedrys, 165
Theobroma beer, 269
Theobroma cacao, 268–69
Thevet, André, 113

thistle, 143
Thorpe, George, 42
thujone, 203
thyme, 323
thymol, 156
Thymus vulgaris, 323
tiswin, 49
Tito's Handmade Vodka, 49
tobacco, 195–96, 197, 236
tobacco liqueur, 195, 196
Toma Verde tomatillo, 347
tomatillo, 347
tomato, 6, 347
Toni-Kola, 313
tonic water, xvii
tonka bean, 197, 242
Top Hat blueberry, 342
torchwood, 249
toxicity, 27–28, 71, 87, 88, 91, 115, 140,
    144, 145, 148, 158, 163, 173, 183, 187,
    192, 194, 195, 197, 208, 223, 233, 256,
    307, 309, 327, 330
trademarks, 228–33
Traminer, 64
trees, 227, 332–39
    see also specific trees
Trigonella foenum-graecum, 160–61
triple sec, 106, 287, 305
triploid, 223
tristeza, 296
Triticum aestivum, 107–10
Tropaeolum majus, 329
true cinnamon, 241–42
Tudor, William, 101
Tungi, 126
turbinado, 100
turmeric, 146
Turnera diffusa, 158
tussock, 92
Tuthilltown Spirits, 50
"twining handedness," 215
tyloses, 53

UCLA, 60
ulanzi, 130
ultraviolet light, 240
Umbellularia californica, 144
umeshu, 279
umqombothi, 49
U.S. Department of Agriculture, 239,
    296
U.S. Department of State, 252

U.S. Patent Office, 92
University of California, Davis, 64
University of Kentucky, 47
University of Minnesota, 39, 67
University of North Carolina, 212
University of Pennsylvania, 32
urak, 114
urushiol, 113

Vaccinium, 342
Valencia orange, 301
Valenzuela-Zapata, Ana, 6
Van Wees Tonka Bean Spirit, 197
Vance, Zebulon, 92
vanilla, 53, 151, 153, 198–99
Vaughan, Benjamin, 257
Vavilov, Nikolai, 25
Vecchio Amaro del Capo liqueur, 314
vegetables, 345–54
    see also specific vegetables
velum, 61
Velvet Falernum liqueur, 102, 137, 298
Veratrum album, 163
verbena, 175–77
Vermont White vodka, 258
Vermont Spirits distillery, 258
vermouth, 63, 65, 140, 153, 173, 205
Vermouth Cocktail, 61
Verveine du Velay liqueur, 177
Vin Mariani wine, 154–55
vines, 340–44
vinho d'batata, 73
vins doux naturels, 62
viola, 329
Viola adorata, 225–26, 329
violet, 225–26, 329
Violette de Bordeaux fig, 334
Virgil, 293
Virginia Land Law, 45
vitamin B, 89
vitamin C, 255, 256, 264
Vitis riparia, 67
    vinifera, 57–69
vodka, 70, 85, 102, 167, 172
    artisanal potatoes and, 74–75
    corn, 49
    garden cocktails with, 348
    infused, 343
    invention of, 72
Vogelbeerschnaps, 281
von Humboldt, Alexander, 228, 232
Voodoo Tiki distillery, 126

THE DRUNKEN BOTANIST

STEWART

Wall Street Journal, 199
Waller, Edmund, 172
walnut, 315–17
Wapsie Valley corn, 50
Wari people, 190
Washington, George, 23–24, 84–85
water, 21
 bourbon and, 47
watermelon, 347
waxy corn, 46
Weber, Franz, 12
Weber, Frédéric Albert Constantin, 12
Weber Blue agave, 10, 11, 12, 15
Wells & Young's Brewing Company, 112
Wells Banana Bread Beer, 112
West Indian lime, 298, 334
West Indian Medical Journal, 248
wheat, 83, 84, 107–10
 lemon wedge and, 109
 nitrogen, protein and, 107–8, 110
 a touch of, 108–10
 varieties of, 110
whiskey
 "American-style," 54
 best way to drink, 40
 bugs in booze, 41
 color of, 33
 "common," 84–85
 malting, 37, 39
 or whisky, 35
 rye, 85–86
 Scotch whisky, 37–40
White, E. B., 165
white currants, 266, 342
white dog, 49
white pine blister rust, 264–65, 342
white spruce, 256
"white whiskey," 23

wild bird pepper, 352
wild spinach, 105
wine, 57–69
 American experiment, 63–67
 aromatized, 63, 65
 brandy, 60–63
 first, 60
 fortified, 60–63
witchweed, 95
Wonderful pomegranate, 335
Woodford Reserve distillery, 50
wormwood, 66, 200–203
 species used in liqueurs, 203
Wucher, Marc, xiii

xanthones, 164
xylitol, 235

yam, 73, 191
Yamada Nishiki rice, 78
yaws, 41
yeast, 26–28, 59, 63, 106, 211
Yellow Field Corn, 50
Yellow Pear tomato, 347
yew, 256
yucheong, 302
yuzu, 302

Zanzibar cloves, 153
Zea mays, 42–51
Zefa Fino fennel, 322
zester, 293
Zingiber officinale, 166–67
Zirbenz Stone Pine Liqueur, 250
Zoco liqueur, 284
Đubrówka, 146, 147
zuppa inglese, 56
Zymomonas mobilis, 3–4